S0-CBG-627

Environmental Mutagenesis, Carcinogenesis, and Plant Biology

Environmental Mutagenesis, Carcinogenesis, and Plant Biology

Volume II

Edited by
Edward J. Klekowski, Jr.

PRAEGER

PRAEGER SPECIAL STUDIES • PRAEGER SCIENTIFIC

Library of Congress Cataloging in Publication Data
Main entry under title:

Environmental mutagenesis, carcinogenesis, and
plant biology.
Bibliography: v. 1, p.; v. 2, p.
Includes indexes.
1. Mutagenesis. 2. Carcinogenesis. 3. Mutagens.
4. Carcinogens. 5. Plants—Metabolism. 6. Plant
metabolites. I. Klekowski, Edward J.
QH465.A1E58 581.2′4 81-21003
ISBN 0-03-061601-8 (set) AACR2
ISBN 0-03-057953-8 (v. 1)
ISBN 0-03-061602-6 (v. 2)

V.2

41,886

Published in 1982 by Praeger Publishers
CBS Educational and Professional Publishing
a Division of CBS Inc.
521 Fifth Avenue, New York, New York 10175 U.S.A.

© 1982 by Praeger Publishers

All rights reserved

23456789 145 987654321

Printed in the United States of America

Contributors

William S. Barnes

Division of
Nutrition Biochemistry
Naylor Dana Institute
for Disease Prevention
American Health Foundation
Valhalla, New York

William F. Grant

Genetics Laboratory
Department of Plant Science
Macdonald Campus of
McGill University
Ste. Anne de Bellevue,
Quebec, Canada

Edward J. Klekowski, Jr.

Botany Department
University of Massachusetts
Amherst, Massachusetts

Robert A. Nilan

Program in Genetics and
Cell Biology and Department
of Agronomy and Soils
Washington State University
Pullman, Washington

Jeffrey L. Rosichan

Program in Genetics and
Cell Biology
Washington State University
Pullman, Washington

R. C. Sautkulis

Biology Department
Brookhaven National Laboratory
Upton, New York

L. A. Schairer

Biology Department
Brookhaven National Laboratory
Upton, New York

Baldev K. Vig

Department of Biology
University of Nevada at Reno
Reno, Nevada

CAMROSE LUTHERAN COLLEGE
LIBRARY

For my daughter Amanda,
her generation will inherit
our environment

Contents

Preface

Concern over human exposure to mutagens (agents causing permanent genetic changes) remained primarily academic until the finding that the majority of chemical carcinogens are also mutagens. Although the converse relationship—that the majority of mutagens are also carcinogens—is not logically necessary, it is prudent to identify and understand environmental mutagens in light of their potential as human carcinogens.

No one can dispute the primary significance of plants in human carcinogenesis. The most common human carcinogen is the smoke produced by burning dried leaves of *Nicotiana tabacum*. Since "all flesh is grass," the presence of naturally occurring mutagens and antimutagens in our food and fodder is of interest. The capacities of plant metabolism to transform promutagens into metabolites that may enter the mammalian food chain take on new significance when one considers the diversity of xenobiotics (pollutants and agrochemicals) to which plants are exposed. Finally, recent researches have shown that plant systems represent some of the most sensitive eukaryotic mutagen bioassays; thus it is important to understand plant promutagen activation and how it parallels mammalian hepatic metabolism. The goal of these volumes is to serve both as a current compendium of plant-related research in environmental mutagenesis and as a catalyst for future studies.

I would like to thank my wife, Libby, for her help in all aspects of this work.

Edward J. Klekowski, Jr.

1

Plant Mutagen Assays Based upon Chromosome Mutations

William F. Grant

INTRODUCTION

Higher plants offer many advantages for the screening and monitoring of environmental chemicals for their potential mutagenicity—that is, for determining whether a chemical can cause an abrupt, heritable change in the composition or arrangement of genes (Nilan and Vig, 1976; de Serres, 1978; Grant and Zura, 1982; Grant, Zinov'eva-Stahevitch, and Zura, 1981). Although many higher plant mutagen assay systems, including that for chromosome aberrations, have been in use for many years, these highly reliable systems have been largely ignored by mammalian toxicologists who are concerned with regulatory policies. As Stich and San (1980) stated, "Whatever the reasons may be, the use of plants in monitoring and screening systems for mutagens has not been widely accepted. The recent successful introduction of the use of *Tradescantia* staminal hairs to detect airborne mutagens and carcinogens may be the beginning of the recognition of various plant assays which are inexpensive, easy to handle and applicable to indoor as well as outdoor detection of environmental mutagens."

Plant chromosome studies have had a strong impact on our understanding of the nature of and the processes affecting the structural integrity of chromosomes. The classical test—the *Allium* test—was developed by Levan (1938; 1949) as an assay system for the analysis of the effects of chemicals such as mitotic poisons on chromosome behavior. Two other species, *Vicia faba* and *Tradescantia paludosa*, have been used extensively in the now-classic studies of Lea, Sax, Dubinin, Gustafsson, Stubbe, Sparrow, and others to characterize the mechanisms and types of chromosome aberrations induced by ionizing radiations (Auerbach, 1976).

In these early studies it was shown that chemicals, such as nitrogen mustard (Novick and Sparrow, 1949), not only caused chromosome aberrations in plants but also produced a spectrum of chromosome aberrations different from that of ionizing radiations (Ford, 1949). Other early studies showed that chromosome irregularities in meiosis could lead to partial sterility in plants that had been sprayed with nicotine sulfate (Kostoff, 1931). Thus chromosomal aberration data were gathered as an index of mutational damage. Such studies eventually led to the idea that chemicals in our environment might be posing a threat to the integrity of the human genetic material. This early information gained from classical plant chromosome studies was available and readily adaptable to the study of environmental chemicals for the detection of mutagens. This review will outline some of the chromosome assays used for detecting the effects of chemical mutagens in plants.

PLANT TEST SYSTEMS

In contrast with the more than 500 species that have been used in mutagenesis studies, relatively few species have been used for determining whether a particular chemical is a clastogen (capable of breaking chromosomes; Shaw, 1970) or a turbagen (capable of causing mitotic disturbances such as C-mitoses; Brøgger, 1978). Ideally, plants with a small number of large, easily identifiable homologous chromosomes are desirable for the study of chromosome aberrations. In some cases—for example, in *Hordeum vulgare* and *Vicia faba*—it has been possible to alter the karyotypes to make the plants more suitable to the investigator's need (Michaelis and Rieger, 1971; Nicoloff and Künzel, 1976). The haploid complement at the time of pollen mitosis offers a unique means of analyzing the extent of mutagen-induced chromosome damage for which there is no parallel in the animal kingdom (Ma et al., 1973; 1978; Smith and Lotfy, 1954).

In an examination of the literature cited by Shelby (1976), a tally was made of the plants most frequently used for the study of root tip chromosome aberrations. It was found that there were 538 references to eight different species (Table 1.1). As may be noted

TABLE 1.1
Plants Most Frequently Used for the Analysis of Chromosome Aberrations (538 studies)

Species	*Common Name*	*No. of Times Used in a Study*	*Percent*
Vicia faba	broad or fava bean	196	36.4
Allium cepa	onion	156	29.0
Hordeum vulgare	barley	94	17.5
Crepis capillaris	hawk's-beard	36	6.7
Pisum sativum	pea	22	4.1
Tradescantia paludosa	spiderwort	13	2.4
Zea mays	corn	12	2.2
Lycopersicon esculentum	tomato	9	1.7

in Table 1.1, *Vicia faba* has been used in over a third of the studies, and is followed closely by *Allium cepa.* Historically, it is interesting to observe that the species *Vicia faba*, *Allium cepa*, *Pisum sativum*, and *Lycopersicon esculentum* were used as far back as 1937 by Marshak to study the effects of X rays on chromosomes.

In addition to the analyses carried out on root tip cells, meiotic end points also may be examined, and comparisons made for both mitotic and meiotic assays from the same treatment. Such studies have most frequently been carried out with *Hordeum*, *Tradescantia*, *Zea*, and *Lycopersicon.*

CYTOLOGICAL END POINTS

Following treatments with a clastogenic chemical, one might expect to observe breaks, exchanges, and gaps that may involve subchromatid, chromatid, and chromosome aberrations and rearrangements. Assays involving clastogenic chemicals have focused primarily on the identification and quantification of unstable aberrations such as dicentrics, acentric fragments, or the products of nonreciprocal translocations. Translocations and inversions can be detected if they involve fairly large proportions of chromosomes.

Metaphase and anaphase scoring are most commonly used for the analyses of induced chromosome aberrations. Chromosome aberrations appearing in anaphase can be scored more readily than those in metaphase (Nicoloff and Gecheff, 1976), and therefore provide a rapid means of estimating chromosome damage. In contrast, metaphase analyses give more information on the types of aberrations that a chemical produces.

Chromosome aberration data can be expressed as the number of a specific type of aberration per 100 cells or the total number of various aberrations per 100 cells.

Some chemicals produce a colchicine (C-mitotic) effect only, or in addition to chromosome aberrations. C-mitosis was first described by Levan (1938) on root tip mitoses of *Allium cepa.* Such chemicals cause an inactivation of the spindle mechanism, leading to restitution nuclei with double the chromosome number. Partial spindle inactivation may lead to irregular anaphase movement of the chromosomes and multipolar anaphases (Wuu and Grant, 1966a). Chemicals may also cause a supercondensation and pycnosis of the chromosomes and/or nuclei at interphase (Mohandas and Grant, 1972), a condensation at prophase and later stages of mitosis, chromosome stickiness, micronuclei in telophase (Ma, 1979), and incomplete furrowing of the cytoplasm (Wuu and Grant, 1966b). Descriptions of the effects of physical and chemical agents on chromosomes of numerous organisms (Evans, 1962; Wolff, 1963, Kihlman, 1966; 1971; Auerbach, 1976), and those resulting specifically from pesticide treatments (Grant, 1978), have been reviewed. Highlights of the various cytogenetic effects produced also have been given in a number of textbooks (Swanson, 1957; Strickberger, 1976).

HIGHER PLANT CHROMOSOME ASSAYS USED FOR THE DETECTION OF CHROMOSOME ABERRATIONS

Root tips are most frequently used for studying chromosome aberrations, but mitotic chromosome aberrations may also be observed in other meristematic cells, such as leaf and stem primordia, or in tapetal cells.

The procedure in root tip chromosome analyses is similar for the most commonly used plants listed in Table 1.1. Detailed protocols have been prepared by the Gene-Tox Committee for plant genetic and cytogenetic assays (Constantin and Owens, 1982). These include root tip assays for chromosome aberrations in *Allium* (Grant, 1982), *Hordeum vulgare* (Constantin and Nilan, 1982), *Tradescantia* (Ma, 1982b), and *Vicia faba* (Ma, 1982a). For specific details concerning procedures for these plants, the respective papers should be consulted. A general schedule will be given for the preparation of root tips for chromosome aberration analyses. For many alternatives and additional data, Sharma and Sharma (1972) or Darlington and La Cour (1976) should be consulted.

Vicia faba Root Tips as an Assay System for the Analyses of Chromosome Aberrations

As may be seen in Table 1.1, chromosome assays for studying the effects of chemicals have been carried out with *Vicia faba* more than any other species. The large chromosomes of this species are ideal for studying the effects of radiation (Read, 1959) and potential chemical mutagens (Grant, 1978). The six pairs of chromosomes are designated according to centromere position as either M (median) or S (subterminal). The single pair of M chromosomes is more than twice the length of the S chromosomes and possesses a large satellite on the short arm. New karyotypes have been developed by means of structural rearrangements so that all of the chromosomes may be differentiated readily (Döbel, Rieger, and Michaelis, 1973; Schubert et al., 1979a). For the Gene-Tox Program, Ma (1982a) summarized the dose-response (positive or negative) of 86 chemicals that have been studied for clastogenicity in *Vicia faba* root tip assays.

The seeds of *Vicia faba* are large and do not lend themselves to germination in a petri dish. Various techniques have been developed for their germination and culture (Kihlman 1971; 1975a). Seeds may be germinated in paper toweling, but to prevent fungal growth, it is best to germinate in aerated, running tap water or in an inert substance such as perlite at 20°C, using a thermostatically controlled heater.

The technique for culture and treatment practiced in our laboratory (developed by K. D. Zura) is as follows:

1. When the primary roots reach approximately 1 cm (3–4 days), the seedlings are planted in a bed of moist perlite. The bed consists of a rectangular Plexiglas frame (16 × 14 cm) with plastic screening secured to the bottom. Perlite is spread over the screening to a depth of 2.5 cm. The seedlings (4 per bed) are planted so that the primary root of each seedling protrudes down through the screen, allowing the seedling to obtain nutrients from the solution below. The bed is placed on top of a rectangular glass container whose exterior surface is painted black.
2. Fill the container to within 0.5 cm of the rim with Hoagland's nutrient solution (Hoagland and Arnon, 1938), in which the iron may be supplied in the form of 1 ml/l of a 0.5 percent $FeSO_4$ solution. Adjust the pH to 6.
3. The root tips of the seedlings are immersed in the nutrient solution, which is aerated continuously with an air pump. Several containers can be maintained continually in controlled environmental conditions: 16-hour photoperiod, 24 ± 1°C light, 19 ± 1°C dark.
4. Decapitate primary roots when they reach a length of approximately 4 cm.
5. Maintain plants in nutrient solution until the secondary roots reach 3–4 cm, at which time they are ready for treatment. Plants used for negative and positive controls are cultured under identical conditions at pH 6. Of the 4 seedlings in each bed, 2 or 3 will reach this stage at the same time. Up to 40 secondary roots may be expected from each seedling.

Other practices to follow in order to standardize the procedure have been documented by Kihlman (1971; 1975a). A protocol for the handling of root tips for fixation, staining, and so on will be given later.

Allium cepa Root Tips as an Assay System for the Analyses of Chromosome Aberrations

Like *Vicia faba*, onions have been used at least since 1937 for chromosome assays to study the effects of X rays on chromosomes (Marshak, 1937). The first major use of onions to study the effects of chemicals on plant chromosomes dates back to the studies of Levan (1938; 1949), who developed what became known as the *Allium* test. Although *A. cepa* has been the most commonly used member of the *Allium* genus in mutagenesis studies, *A. carinatum*, *A. fistulosum*, and *A. sativum* have also been used (Grant, 1982). All of these species have a somatic chromosome number of 16.

The karyotype of *A. cepa* consists of eight pairs of chromosomes (2n = 16): one pair of satellite chromosomes, two pairs submedian, and five pairs median to submedian. The duration of the mitotic cycle has been determined by several authors to be about 20 hours. Matagne (1968) has reported the duration of the interphase periods G_1, S, and G_2 to be 3.3, 12, and 3.7 hours, respectively, and mitosis 4 hours at 20°C.

In contrast with *Vicia faba*, where secondary root tips of seedlings are treated, root tips from growing onion bulbs are treated directly. A primary difference, however, is in the initial selection and handling of the bulbs for root growth. Commercial bulbs of onion may be treated with a growth inhibitor, such as maleic hydrazide, to prevent sprouting during storage. Such bulbs may produce poor and uneven root growth, and should not be used. Young bulbs of relatively uniform size (15–30 g in weight) should be selected so that the base will fit a round-mouth jar or holes in a rack. The bulbs should be denuded before use by removing the loose outer scales, and the base scraped so that the apices of the root primordia are exposed. The bulbs should be placed so that their bases remain submerged in tap water. The water is changed daily or continuously aerated. Treatment may begin when the roots reach a length of 15 to 20 mm. The root tips should reach this length in approximately two to four days. The bulbs are grown in the dark for the period of the experiment. The containers may be wrapped in heavy black paper or aluminum foil to give added protection from the light.

The treatment and further handling of the root tips in preparation for chromosome aberration analyses are the same as for the small-seeded species, which will be discussed in the next section.

Schedule for the Treatment of Small-Seeded Species

Species that produce small seeds can all be handled in a similar manner for germination and treatment. With the exception of *Tradescantia paludosa* (to be discussed later) and *Vicia faba* and *Allium cepa* (already treated), the remaining species listed in Table 1.1 that have been used very frequently for chromosome aberrations studies—*Hordeum vulgare*, *Crepis capillaris*, *Pisum sativum*, *Zea mays*, and *Lycopersicon esculentum*—fall into this category.

Seed Treatment

It may be necessary to surface-sterilize seeds for mold. This may be done by soaking the seeds for 15 to 30 minutes in 1 to 5 percent calcium hypochlorite solution. Then drain and rinse thoroughly in several changes of distilled water. Dry seeds are soaked in freshly prepared solutions of the chemical to be tested. In most cases suitable concentrations and treatment periods (concentration × time) will have to be determined for the target cells. Freshly prepared solutions of the chemical to be tested should be used, since many chemicals are unstable in solution. In the author's laboratory, concentrations of 500, 1,000, and 1,500 ppm for periods of 6, 12, and 24 hours have been used (Wuu and Grant, 1966a). In some cases the chemicals have been too toxic, completely inhibiting germination. In such cases, concentrations have been reduced (100, 200, and 300 ppm) and/or the time periods shortened (two, four, and six hours). Treatment solutions should be maintained at or near neutral pH. Solutions using a 0.1 M monobasic and dibasic sodium phosphate buffer may be employed to maintain pH.

Treatment of Seedlings

In asexually propagated plants, such as clones of *Tradescantia* that are heterozygous for flower color, a large number of root tips may be obtained from plants that are propagated by means of cuttings. Cuttings may be planted in flats filled with vermiculite and provided with bottom heat. They can be kept moist by means of an automatic spray mist regulated by a moisture sensor. In the case of *Tradescantia*, root tips ready for treatment will develop in 20 days (Ahmed and Grant, 1972). Root tips are washed prior to placing them in beakers containing the treatment solution. The top and sides of each beaker are covered with heavy black paper and the beakers are aerated by means of an air pump.

Schedule for Root Tips

There are many variations in schedules for the preparation of root tips (fixation, staining, maceration, and so on) for cytological examination (Sharma and Sharma, 1972; Darlington and La Cour, 1976). A general procedure used in the author's laboratory will be given.

1. *Recovery Period.* Following treatment, the root tips should be allowed to recover in water for one complete cell cycle, since chromosome aberrations may be produced by different chemicals at different stages in the interphase period (G_1, S, or G_2). Thus one may determine whether primarily chromosome or chromatid aberrations are produced.

For the species that have been most frequently used for chromosome aberration analyses, the duration of the cell cycle has been determined, but for particular growing conditions or species not previously studied, the length of the cell cycle should be determined. A few techniques that have been used are discussed below.

Gonzalez-Fernandez, Lopez-Saez, and Gimenez-Martin (1966) developed a method for producing binucleate cells in *Allium cepa* that can be used for determining the length of the cell cycle. The rationale is to inhibit cytokinesis, with the result that two nuclei are formed in the same cell (binucleate cell). The two nuclei in a binucleate cell undergo simultaneous mitosis (bimitosis). The duration of the binucleate cell divisions can be determined fairly precisely.

To produce binucleate cells, Gonzalez-Fernandez and colleagues treated onion root tips with a 0.1 percent caffeine solution for 1 hour, after which they were washed and reimmersed in tap water. After 1 hour recovery an average of 4.26 percent (range 3.63 to 5.04 percent) of the cells were binucleate in an analysis of 2,000 to 3,000 cells. After 10 hours recovery the binucleate cells began the division process. After 23 hours a new series of cells began to undergo division (second-order binucleate cells), reaching a maximum after 28 hours. Thus, a complete division cycle lasts 14 hours. The authors showed that the division cycle of the bimitotic cells closely parallels the time of the division cycle of the mononucleate cells, which lasted 13.5 hours. The authors consider that the advantage of using caffeine is that it presents a very low toxicity to the cell.

The mitotic cycle time can also be estimated from the rhythmic appearance and disappearance of the number of labeled cells in mitosis with time, and is generally scored as the interval of time between two successive ascending portions of such a curve. Van't Hof (1965) has used both colchicine and ^{3}H-thymidine to determine the length of the cell cycle for *Pisum sativum*. Cells entering mitosis are affected by colchicine in such a manner as to prevent cytokinesis, but not karyokinesis. Consequently, the cells become tetraploid in subsequent mitoses, and can be used as a measure of the duration of the mitotic cycle. Cells can be labeled with ^{3}H-thymidine when DNA synthesis occurs in interphase.

Van't Hof (1965) considers the duration of the cell cycle to be the period that elapses between the treatment with colchicine and the time when the maximum number of tetraploid cells is observed. He has found this period to be 14 hours for colchicine, with mitosis averaging 1.83 hours.

2. *Pretreatment.* After the recovery period, harvest the root tips, and, depending on the experiment, either pretreat prior to fixation or fix immediately. For pretreatment a 0.05 to 0.1 percent aqueous solution of colchicine is most commonly used to obtain the large number of metaphases that are required for the scoring of interchanges and deletions. At 20°C a pretreatment period of one to two hours should suffice. An alternative to colchicine is to use a 0.002 M 8-hydroxyquinoline solution with a treatment period of one to two hours. If only anaphase aberrations are scored, pretreatment should be omitted, since colchicine and hydroxyquinoline destroy the spindle, thereby preventing the mitotic cycle from continuing beyond metaphase.
3. *Fixation.* Fix in absolute or 95 percent alcohol-glacial acetic acid (3:1) for a convenient period ranging from 30 minutes to 24 hours. After fixation the material can be maintained in 70 percent alcohol in a refrigerator, and the procedure resumed at the next step when convenient.
4. *Staining.* The usual procedures for aceto-carmine, aceto-orcein, propionic-carmine, proprionic-orcein, or Feulgen reagent may be followed. The latter procedure is specific for DNA, and is generally preferred because of the excellent contrast obtained. The procedure is as follows: hydrolyze in 1 N HCl at 60°C for about 5 minutes (optimum time, between 4 and 12 minutes; will vary with specific conditions), and stain in Feulgen reagent for 2 hours.
5. *Maceration.* Treat with 5 percent pectinase for one to three hours. If roots are left too long in pectinase, they will become overly soft and difficult to handle.
6. *Slide Preparations.* On a slide remove the meristematic region with a dissecting needle in a drop of 45 percent acetic acid, and squash the material with a cover slip. Better spreads can be obtained by gently tapping the cover slip with a dissecting needle before squashing.
7. *Temporary Sealing.* Seal the cover slip with a paraffin-gum arabic mixture, clear fingernail polish, or rubber cement to provide a temporary mount.
8. *Controls.* Set up control experiments in which all the above steps are followed. A positive control should be used for comparative purposes, selecting an agent such as ethyl methanesulfonate that is known to induce chromosome aberrations. A solvent control should be used if a solvent other than water is required to prepare a solution of the test chemical. In addition, a water control and a buffer control should be used. Variation in type and frequency of chromosome aberrations between the controls and test chemical will give a measure of the activity of the compound.
9. *Permanent Slides.* A temporary slide will deteriorate after a few days, even when kept in a petri dish and stored in a refrigerator. To make it

permanent, use dry ice (Conger and Fairchild, 1953), liquid carbon dioxide, or a freezing apparatus to lower the temperature below -40°C (Sharma and Sharma, 1972). When the preparation is frozen, the coverslip is removed with a razor blade or scalpel. The root tip material adheres to the slide, which subsequently is immersed in two changes of absolute alcohol and mounted in Euparal with a clean cover slip.

Sister Chromatid Exchanges

In the last few years sister chromatid exchanges (SCEs) have become important as an assay in evaluating chromosome damage (Perry and Evans, 1975). It is considered that their measurement is the most sensitive mammalian cell technique for detecting mutagenic carcinogens (Wolff, 1979).

SCEs were first demonstrated by means of autoradiography as switches in a tritiated thymidine label from one chromatid to its sister (Taylor, Woods, and Hughes, 1957; for review, see Kato, 1977). The relatively poor resolution of autoradiography made an accurate estimate of the number of exchanges unreliable. With the development of new techniques, the use of SCEs in the testing of mutagens has provided a rapid, sensitive, and quantitative assay for genetic damage. A predictable and relatively stable "background level" of SCEs is found after normal preparation. These may be induced during preparation, but it has been amply demonstrated that the frequency of SCEs can be increased by chemical mutagens and at concentrations considerably lower than those required to produce chromosomal aberrations (Kihlman and Sturelid, 1978).

It has been shown that within species, SCE formation is positively correlated with chromosome length both in plants (Kihlman and Kronborg, 1975; Schubert et al., 1979a) and in animals (Tease and Jones, 1979), and that a tetrapoloid exhibits twice as many SCEs per cell as its diploid cytotype (Friebe, 1980). A study by Carrano et al. (1978) has demonstrated a linear relationship between induced SCEs and mutation, indicating that SCEs are closely associated with mutational damage. While considerable information is available on the chemistry involved in the differentiation of sister chromatids, their biological significance is not known (Evans, 1977). The relationships between the mechanisms involved in the function of chromosomal aberrations and SCEs are still controversial, ranging from related to completely different (Schvartzman and Hernandez, 1980).

The technique that has been most successful in demonstrating

SCEs is the fluorescence plus Giemsa (FPG) technique developed by Perry and Wolff (1974), and first used by Kihlman and Kronborg (1975) to demonstrate SCEs in *Vicia faba*. The technique has been referred to as a BrdU-labeling method, since 5-bromodeoxyuridine is substituted for thymidine in the chromosomal DNA (Kato, 1977). Briefly, the principle of the FPG or BrdU-labeling technique is as follows: A chromatid that has incorporated BrdU into both strands of its DNA stains more weakly with the FPG technique than does a chromatid that has incorporated BrdU into only one strand. Such a unifilarly substituted chromatid does not stain as strongly as a completely unsubstituted chromatid (containing only thymine). After one cell cycle in the presence of BrdU, the two sister chromatids cannot be distinguished by FPG differential staining and the SCEs that may have occurred cannot be detected. However, if the BrdU is omitted during the next cell cycle, one of the chromatids will contain thymine in both DNA strands, whereas the second will be unifilarly substituted with BrdU. Alternatively, if BrdU is supplied during the second cycle, one of the sister chromatids will become bifilarly substituted, whereas the second will remain unifilarly substituted. Under these circumstances sister chromatids can be differentiated by using fluorochrome staining (for instance, "33258 Hoechst") and observed under ultraviolet light, or by the FPG technique. The FPG technique gives a much clearer, permanent resolution of the exchanges than does fluorescence alone, which fades rapidly.

Sister chromatid exchanges in plant chromosomes were first demonstrated by Kihlman and Kronborg (1975) in *Vicia faba*, and later by Kihlman (1975b), Scheid (1976), Vosa (1976), Kihlman, Andersson, and Natarajan (1977), Kihlman, Natarajan, and Andersson (1978), Kihlman and Sturelid (1978), and Schubert et al. (1979a; 1979b). SCEs have been reported in three other species: *Allium cepa* (Schvartzman and Cortés, 1977), *Secale cereale* (Friebe, 1978), and *Hordeum vulgare* (Schubert et al., 1980).

Vicia faba

The procedure described for SCEs for *Vicia faba* by Kihlman and Kronborg (1975), with later modification by Kihlman and colleagues (Kihlman, 1975b; Kihlman, Natarajan, and Andersson, 1978; Kihlman and Sturelid, 1978), is as follows:

1. Lateral roots are exposed for 16-18 hours to an aqueous solution of 100 μM 5-bromodeoxyuridine (BrdU), 0.1 μM 5-fluorodeoxyuridine (FdUrd),

and 5 μM uridine (Urd). Schubert et al. (1979a) expose for 17 hours. (Procedures for the growth of root tips were outlined above. In testing for the cytogenetic effects of a chemical, Kihlman and Sturelid [1978] use four beans for each treatment. Roots of one seedling are used to determine the frequency of chromosomal aberrations. The roots of the other three seedlings are used for scoring the frequency of SCEs.)

2. Replace BrdU solution with one containing 100 μM thymidine (dThd) and 5 μM Urd for 21 hours (another cell cycle) (see Schubert et al. [1979a], 19 hours).
3. Pretreat with 0.05 percent colchicine for three hours (see Schubert et al. [1979a], 2 hours).
4. Fix overnight in cold 1:3 acetic-methanol (see Schubert et al. [1979a], 1 hour). Note: Do not expose roots to light during steps 1 to 4. Treatments should be carried out in the dark at 20°C. Kihlman and Kronborg (1975) used glass tubes (containing 40 ml of solution) covered on the outside with black tape; the main root passes through a small hole in a black rubber disk resting on the top of the tubes.
5. After fixation, rinse roots in 0.01 M citric acid ($C_6H_8O_7 \cdot H_2O$)-sodium citrate ($Na_3C_6H_5O_7 \cdot 2H_2O$) buffer, pH 4.7.
6. Incubate roots for 2 hours at 27°C with 0.5 percent pectinase dissolved in the same buffer. Schubert et al. (1979a) digest with pectinase (1 percent for 2 hours) and cellulase (2 percent for 90 minutes) at 37°C.
7. Squash root tips in 45 percent acetic acid on alcohol-cleaned slides, coated (subbed) with a 10:1 mixture of gelatin and chrome alum (chromium potassium sulfate).
8. Using dry ice (Conger and Fairchild, 1953), liquid carbon dioxide, or a freezing apparatus (Sharma and Sharma, 1972), lower the temperature to below -40°C, freeze preparation, and remove cover slip.
9. Process slide through ethanol series from absolute, 95, 85, 70, 50, and 30 percent to distilled water.
10. Place slide in a moist chamber and incubate for 60 minutes at 27°C (Schubert et al. [1979a], 37°C) in an RNase solution prepared as follows: 1 mg RNase (Sigma RNase-A from bovine pancreas) dissolved in 10 ml 0.5 $\times$ SSC (0.075 M NaCl + 0.0075 M $Na_3C_6H_5O_7 \cdot 2H_2O$). Then 200 μl is placed onto the tissue and the preparation is covered with a fresh cover slip. The RNase step is not absolutely necessary, but it improves the contrast between sister chromatids. Schubert et al. (1979a) then dehydrate the slide in ethanol and air-dry it before proceeding to step 11.
11. Rinse in 0.5 $\times$ SSC.
12. Stain for 20 minutes with a solution of "33258 Hoechst" prepared as follows: 1 mg of the fluorochrome dissolved in 1 ml ethanol and 0.1 ml of this solution added to 200 ml 0.5 $\times$ SSC.
13. After staining, rinse the preparations and mount them in 0.5 $\times$ SSC; temporarily seal cover slips with gum mastic, clear fingernail polish, or rubber cement.
14. Observe preparation, using a fluorescence microscope. With a Zeiss ultraviolet microscope, exciter-filter combinations BG38 and BG12 and barrier

filters 43 or 54 may be used. Differentiation of sister chromatids may be improved by exposing the slides to long-wave ultraviolet light for 60 minutes at a distance of 6 cm from the light source.

15. Store slides over distilled water in a moist chamber for four days at 4°C and then incubate for 60 minutes at 55°C in 0.5 × SSC. The latter step may be omitted if the slides are stored at room temperature (20°C) under ordinary light conditions for one day, and for an additional period of three days in the dark in a refrigerator at 4°C.
16. Rinse slides thoroughly in 0.017 M phosphate buffer, pH 6.8.
17. Stain for 7 minutes in a solution containing 3 percent Giemsa (Gurr's R66) in 0.017 M phosphate buffer, pH 6.8.
18. Rinse preparations, first in phosphate buffer, then in distilled water, and air-dry them.
19. Dip the dry preparations in xylene and mount in a mounting medium such as Permount, Eukitt, Euparal, or Canada balsam.
    For steps 12 to 17, Schubert et al. (1979a) stain for 45 minutes under ultraviolet irradiation, using a quartz lamp. Following a short rinse in 0.5 × SSC, the slide is incubated (step 15) in 2 × SSC at 80°C for 1 hour. For step 17, after a quick rinse in 0.5 × SSC, stain in Merck-Giemsa (diluted 1:25 in 0.067M Sörensen's phosphate buffer, pH 6.7) for 2 minutes.

For the testing of a putative mutagen, root tips of *Vicia faba* are exposed for one or two hours between steps 1 and 2.

Allium cepa

The procedure of Schvartzman (1979), Schvartzman and Cortés (1977), Schvartzman et al. (1979a; 1979b), and Schvartzman and Hernandez (1980) for SCEs in *Allium cepa* is as follows:

1. Root tips of bulbs are exposed to a treatment solution containing 100 μM BrdU, 0.05 to 0.001 μM FdUrd, and 1.0 μM Urd for 21 hours in order to ensure that cells incorporate BrdU throughout one S period. Schvartzman, Postigo, and Gutiérrez (1979b) state that variations in FdUrd concentration up to 5×10^{-8}M (0.05 μM) do not significantly modify either the cell cycle duration or the yield of the SCEs.
2. Replace with a solution containing 100 μM thymidine (Thd) and 1.0 μM Urd for 14 hours ensuring Thd incorporation for a second S period.
3. Pretreat root tips with 0.05 percent colchicine for three hours.
4. Fix in methanol-acetic acid (3:1) at 4°C overnight. Note: The culture receptacles and the treated bulbs were kept in darkness.
5. After fixation, treat root tips for one hour with 0.05 percent pectinase dissolved in citrate buffer adjusted to pH 4.2 at 37°C, and then squash them.
6. Remove cover slip and dehydrate root tips by passing them through ethanol (100 percent, 96 percent, 70 percent, 50 percent, 30 percent) to distilled water.

7. Incubate with RNase at 25°C for two hours.
8. Wash slides with 0.5 × SSC.
9. Treat with "33258 Hoechst" for half an hour at room temperature. Preparation of Hoechst: Dissolve 1 mg of fluorochrome in 1 ml of ethanol; add 0.2 ml of this solution to 100 ml of 0.5 × SSC.
10. Wash slides with McIlvaine buffer adjusted to pH 8.0.
11. Apply cover slips and seal them with nail polish.
12. Expose slides to a fluorescent sun lamp radiating in the 280-380 nm band (Westinghouse FS 20) at 10 cm distance on a hot plate at 30°C for one hour.
13. Remove cover slips and treat slides at 55°C with McIlvaine buffer for one hour.
14. Wash several times with phosphate buffer adjusted to pH 6.8 (concentration of phosphate buffer not given; presumably 0.01 M after Goto et al., 1978).
15. Stain slides with 3 percent Giemsa (Gurr's R66) in phosphate buffer adjusted to pH 6.8 for nine minutes.
16. Wash in same buffer.
17. Air-dry and mount with Euparal.

Hordeum vulgare

Sister chromatid exchanges also have been successfully induced in barley (Schubert et al., 1980). The procedure is as follows:

1. Treat roots of germinated seeds with a solution of 5×10^{-4} BrdU, 5×10^{-8} M FdUrd, and 10^{-6} M Urd in darkness at 24°C for 12 hours.
2. Replace with a solution containing 2.5×10^{-5} M Thd and 10^{-6} M Urd for 12 hours.
3. Pretreat root tips with 0.05 percent colchicine for three hours.
4. Fix in ethanol-acetic acid.

Further processing of the material was carried out as described for *Vicia faba* by Schubert et al. (1979a).

Secale cereale

Friebe (1978; 1980) has given his procedure for scoring SCEs in rye as follows:

1. Root tips are treated with a solution of 100 μM BrdU, 0.05 μM FdUrd, and 1.0 μM Urd dissolved in tap water in the dark for 24 hours at 25°C. This corresponds to two cell cycles.
2. Root tips are transferred to 0.05 percent colchicine for three hours.
3. Fix in ethanol-acetic acid (3:1).
4. Treat root tips for two hours in snail-stomach enzyme, then wash.

5. Squash root tips in 45 percent acetic acid and remove cover slip by the dry-ice method.
6. Hydrate through an alcohol series to distilled water.
7. Treat for one hour with a 0.01 percent RNase solution in 0.5 × SSC.
8. After rinsing with 0.5 × SSC, treat the preparation for half an hour with "33258 Hoechst."
9. Rinse preparation in 0.5 × SSC and treat for one hour in 0.5 × SSC under ultraviolet light.
10. Incubate one hour in 0.5 × SSC at 55°C.
11. Rinse in 0.5 × SSC.
12. Stain for 21 minutes in a 3 percent Giemsa solution in M/15 Sörensen phosphate buffer, pH 6.8.
13. Rinse preparations in M/15 Sörensen phosphate buffer, pH 6.8.
14. Air-dry preparations, place them in xylol overnight, and mount them in Euparal.

CHROMOSOME ABERRATIONS IN MEIOSIS AND THE MICRONUCLEUS TEST

Meiotic chromosome aberrations can be studied in all of the species listed in Table 1.1. In contrast with root tips, there have been far fewer studies using meiotic chromosome analyses. For example, Ma (1982a) lists only five studies in *Vicia faba* in which meiotic studies have been carried out as tests for mutagenicity, and only a single meiotic study has been reported in *Allium* (Grant, 1982).

Cells replicating their DNA are approximately five times more sensitive than cells in G_1 or G_2 (Brewen and Preston, 1973); hence, there is generally a higher number of aberrations observed in meiotic cells than in root tip cells from equivalent treatments.

In most species, microsporogenesis does not take place synchronously in all parts of the floral shoot. The exact size of the bud undergoing meiosis can be readily determined with some practice. In *Tradescantia* there is a high degree of synchrony of the anthers within a bud. The general procedures for the preparation of meiotic material for chromosome analyses follow closely those for root tips.

Ma et al. (1978) and Ma (1979) state that the meiotic chromosomes in *Tradescantia* cannot be clearly stained when fixed, and recommend the use of the micronucleus test as an indication of chromosomal damage for environmental mutagen testing as well as for in situ monitoring. Micronuclei are usually formed in the quartet stage from acentric fragments or clumped chromosomes that have not been incorporated into the restitution nuclei. Micronuclei may be readily stained using the aceto-carmine squash method. The scoring of 400 quartet cells will indicate the degree of chromosome damage. Ma et al. (1981) report a clone of *Tradescantia reflexa* that is two to ten times more sensitive to mutagens than *T. paludosa* in the micronucleus assay.

SOMATIC CHROMOSOME ABERRATIONS IN MICROSPORES AND POLLEN TUBES

Mitosis in the male gametophyte of *Tradescantia* has been used to monitor the effect of chemicals and radiation on chromosomes (Ma and Kahn, 1976; Rushton, 1969; Smith and Lotfy, 1954). Somatic divisions of either the microspore or the generative nuclei are scored. The study of the microspore nuclei has the advantage that, in the first postmeiotic division, the microspores undergo a relatively synchronous division, thus permitting examination of a large population of cells at the same stage.

To test for chromosome aberrations in generative nuclei in response to chemical agents, Smith and Lotfy (1954) carried out the following variation of the coated slide method:

1. Harvest fresh pollen in the morning and place it in a desiccator at 6°C for four hours.
2. Treat dried pollen with chemical to be tested.
3. Sow immediately on culture medium (1 percent agar, 12 percent lactose, and 0.01 percent colchicine) that has been spread lightly on a cover glass to form a thin film.
4. Invert cover glass over a glass ring attached to a slide to form a Van Tieghem cell that may be humidified with a small piece of moist absorbent paper.
5. Place cultures in an incubator at 25°C for 24 hours. A number of metaphases of the dividing generative nuclei should be present; controls reach metaphase on an average of two to four hours sooner.
6. Remove cover slips and place in acetic-alcohol (1:3) for a few seconds.
7. Invert cover slip in a large drop of aceto-carmine on a slide for several minutes.
8. Remove cover slip with a pair of tweezers.
9. Pass through the following dehydration series:
    a. 70 percent alcohol:glacial acetic acid (3:1)
    b. absolute alcohol:glacial acetic acid (3:1)
    c. absolute alcohol:glacial acetic acid (9:1)
    d. absolute alcohol.

    A prolonged immersion (more than 15 seconds) in absolute alcohol may cause the development of crystals in the medium that makes examination of the slide difficult.

10. Mount slide in Euparal or other medium, and examine microscopically.

Alternatively, the pollen can be cultured by using the hanging drop method (Sharma and Sharma, 1972) or the floating cellophane technique (La Cour and Fabergé, 1943). The culture medium can be varied by substituting sucrose (3-30 percent solution) for lactose or gelatin (1-3 percent) for agar. Extracts from the style, placenta, or stigma have also been added to the medium (Sharma and Sharma, 1972).

RELEVANCE OF CHROMOSOME ABERRATION DATA FROM HIGHER PLANTS TO THOSE FROM MAMMALIAN CELLS

The author has shown that chromosome aberration data from higher plants are comparable with those found for chromosome aberrations in mammalian cells, and therefore has recommended the use of such data from higher plants as an appropriate first-tier assay system (Grant, 1978). In a comparison of some pesticides, excellent correlations have been found between the frequency of both chromosomal abnormalities and C-mitoses in plant and animal systems (Grant, 1978). A similar conclusion was drawn in studies on the effect of eight chemicals in several systems, including in vitro and in vivo mammalian systems, bacteria, *Drosophila*, and plant systems. The response of the plant root tip assay compared favorably with the response of the mammalian cells in culture (Constantin and Owens, 1982). These results indicate that the data from higher plant assays compare as well with those obtained from mammalian in vivo and in vitro assays as do *Drosophila* or bacteria. Therefore, a higher plant chromosome assay should be included as one recommended assay in the battery of tests used for determining the mutagenicity of environmental chemicals.

GENERAL COMMENTS

Higher plant chromosome assay systems are simple, reliable, and inexpensive. Numerous studies have demonstrated that plant chromosome assays are sensitive indicators of environmental chemicals. Chromosome breaks should be considered as a warning signal of heritable genetic damage (Nichols, 1973), although some breaks per se may be repaired or lead to the death of a cell line, and therefore do not constitute heritable mutations. Data from higher plant chromosome assays appear to compare favorably with those from mammalian cell cultures. It is recommended that higher plant chromosome assay systems be given their due recognition by committees concerned with regulatory policies.

ACKNOWLEDGMENTS

The technical assistance of Mr. K. D. Zura of my laboratory has been greatly appreciated. Financial support from the National Sciences and Engineering Research Council of Canada for studies of genetic toxicity of environmental chemicals is gratefully acknowledged.

REFERENCES

Ahmed, M., and W. F. Grant. 1972. "Cytological Effects of the Pesticides Phosdrin and Bladex on *Tradescantia* and *Vicia faba*." *Can. J. Genet. Cytol.* 14:157-165.

Auerbach, C. 1976. *Mutation Research: Problems, Results and Perspectives.* London: Chapman and Hall.

Brewen, J. G., and R. J. Preston. 1973. "Chromosome Aberrations as a Measure of Mutagenesis: Comparisons *in Vitro* and *in Vivo* and in Somatic and Germ Cells." *Environ. Health Perspect.* 6:157-166.

Brøgger, A. 1978. "Chromosome Damage in Human Mitotic Cells After *in Vivo* and *in Vitro* Exposure to Mutagens." In *Expert Conference on Genetic Damage in Man Caused by Environmental Agents.* New York: Academic Press.

Carrano, A. V., L. H. Thompson, P. A. Lindl, and J. L. Minkler. 1978. "Sister Chromatid Exchange as an Indicator of Mutagenesis." *Nature* (London) 271:551-553.

Conger, A. D., and L. M. Fairchild. 1953. "A Quick-Freeze Method for Making Smear Slides Permanent." *Stain Technol.* 28:281-283.

Constantin, M. J., and R. A. Nilan. 1982. "Chromosome Aberration Assays in Barley (*Hordeum vulgare*)." *Mutat. Res.* (in press).

Constantin, M. J., and E. T. Owens. 1982. "Plant Genetic and Cytogenetic Assays: Introduction and Perspectives." *Mutat. Res.* (in press).

Darlington, C. D., and L. F. La Cour. 1976. *The Handling of Chromosomes.* 6th ed. London: George Allen and Unwin.

de Serres, F. J. 1978. "Introduction: Utilization of Higher Plant Systems as Monitors of Environmental Mutagens." *Environ. Health Perspect.* 27: 3-6.

Döbel, P., R. Rieger, and A. Michaelis. 1973. "The Giemsa Banding Patterns of the Standard and Four Reconstructed Karyotypes of *Vicia faba*." *Chromosoma* 43:409-422.

Evans, H. J. 1977. "What Are Sister Chromatid Exchanges?" In *Chromosomes Today*, vol. 6, ed. A. de la Chapelle and M. Sorsa, pp. 315-326. Amsterdam: Elsevier/North Holland.

——. 1962. "Chromosome Aberrations Induced by Ionizing Radiation." *Int. Rev. Cytol.* 13:221-231.

Ford, C. E. 1949. "Chromosome Breakage in Nitrogen Mustard Treated *Vicia faba* Root Tip Cells." Proc. 8th Int. Cong. Genet., Stockholm. *Hereditas* (Suppl.): 570-571.

Friebe, B. 1980. "Comparison of Sister Chromatid Exchange and Chiasma Formation in the Genus *Secale*." *Microsc. Acta* 83:103-110.

——. 1978. "Untersuchungen zum Schwesterchromatidenaustausch bei *Secale cereale*." *Microsc. Acta* 81:159-165.

Gonzalez-Fernandez, A., J. F. Lopez-Saez, and G. Gimenez-Martin. 1966. "Duration of the Division Cycle in Binucleate and Mononucleate Cells." *Exp. Cell Res.* 43:255-267.

Goto, K., S. Maeda, Y. Kano, and T. Sugiyama. 1978. "Factors Involved in Differential Giemsa-Staining of Sister Chromatids." *Chromosoma* 66: 351-359.

Grant, W. F. 1982. "Chromosome Aberration Assays in *Allium*." *Mutat. Res.* (in press).

——. 1978. "Chromosome Aberrations in Plants as a Monitoring System." *Environ. Health Perspect.* 27:37-43.

Grant, W. F., and K. D. Zura. 1982. "Plants as Sensitive *in Situ* Detectors of Atmospheric Mutagens." In *Genetic Toxicology: The Utilization of Assays for Mutagenicity*, ed. by J. A. Heddle. New York: Academic Press (in press).

Grant, W. F., A. E. Zinov'eva-Stahevitch, and K. D. Zura. 1981. "Plant Genetic Test Systems for the Detection of Chemical Mutagens." In *Short Term Tests for Chemical Carcinogens*, ed. by H. F. Stich and R. H. C. San. New York: Springer Verlag, pp. 200-216.

Hoagland, D. R., and D. I. Arnon. 1938. "The Water-Culture Method for Growing Plants Without Soil." *Univ. Calif. Agric. Exp. Stn. Berkeley Circ.* no. 347:1-39.

Kato, H. 1977. "Spontaneous and Induced Sister Chromatid Exchanges as Revealed by the BUdR-Labeling Method." *Int. Rev. Cytol.* 49:55-97.

Kihlman, B. A. 1975a. "Root Tips of *Vicia faba* for the Study of the Induction of Chromosomal Aberrations." *Mutat. Res.* 31:401-412.

——. 1975b. "Sister Chromatid Exchanges in *Vicia faba*. II. Effects of Thiotepa, Caffeine and 8-Ethoxycaffeine on the Frequency of SCE's." *Chromosoma* 51:11-18.

——. 1971. "Root Tips for Studying the Effects of Chemicals on Chromosomes." In *Chemical Mutagens*, ed. by A. Hollaender, pp. 489-514. New York: Plenum Press.

——. 1966. *Actions of Chemicals on Dividing Cells.* Englewood Cliffs, N.J.: Prentice-Hall.

Kihlman, B. A., and D. Kronborg. 1975. "Sister Chromatid Exchanges in *Vicia faba*. I. Demonstration by a Modified Fluorescent plus Giemsa (FPG) Technique." *Chromosoma* 51:1-10.

Kihlman, B. A., and S. Sturelid. 1978. "Effects of Caffeine on the Frequencies of Chromosomal Aberrations and Sister Chromatid Exchanges Induced by Chemical Mutagens in Root Tips of *Vicia faba.*" *Hereditas* 88: 35-41.

Kihlman, B. A., H. C. Andersson, and A. T. Natarajan. 1977. "Molecular Mechanisms in the Production of Chromosomal Aberrations. Studies with the 5-bromodeoxyuridine-Labelling Method." In *Chromosomes Today,* vol. 6, ed. A. de la Chapelle and M. Sorsa, pp. 287-296. Amsterdam: Elsevier/North Holland.

Kihlman, B. A., A. T. Natarajan, and H. C. Andersson. 1978. "Use of the 5-bromodeoxyuridine-Labelling Technique for Exploring Mechanisms Involved in the Formation of Chromosomal Aberrations. I. G_2 Experiments with Root-Tips of *Vicia faba.*" *Mutat. Res.* 52:181-198.

Kostoff, D. 1931. "Heteroploidy in *Nicotiana tabacum* and *Solanum melongena* Caused by Fumigation with Nicotine Sulfate." *Bull. Soc. Bot. Bulgar.* 4:87-92; 1934. *Biol. Abstr.* 8:10.

La Cour, L., and A. C. Fabergé. 1943. "The Use of Cellophane in Pollen Tube Technique." *Stain Technol.* 18:196.

Levan, A. 1949. "The Influence on Chromosomes and Mitosis of Chemicals, as Studied by the *Allium* Test." Proc. 8th Int. Cong. Genet. Stockholm. *Hereditas* (Suppl.): 325-337.

——. 1938. "The Effect of Colchicine on Root Mitoses of *Allium.*" *Hereditas* 24:471-486.

Ma, T.-H. 1982a. "*Vicia* Cytogenetic Tests for Environmental Mutagens." *Mutat. Res.* (in press).

——. 1982b. "*Tradescantia* Cytogenetic Tests for Environmental Mutagens." *Mutat. Res.* (in press).

——. 1979. "Micronuclei Induced by X-rays and Chemical Mutagens in Meiotic Pollen Mother Cells of *Tradescantia.* A Promising Mutagen Test System." *Mutat. Res.* 64:307-313.

Ma, T.-H., and S. H. Khan. 1976. "Pollen Mitosis and Pollen Tube Growth Inhibition by SO_2 in Cultured Pollen Tubes of *Tradescantia.*" *Environ. Res.* 12:144-149.

Ma, T.-H., T. Fang, J. Ho, D. Chen, R. Zhou, G. Lin, J. Dai, and J. Li. 1981. "Hypersensitivity of *Tradescantia reflexa* Meiotic Chromosomes to Mutagens." *Mutat. Res.* (in press).

Ma, T.-H., D. Isbandi, S. H. Khan, and Y.-S. Tseng. 1973. "Low Level of SO_2 Enhanced Chromatid Aberrations in *Tradescantia* Pollen Tubes and Seasonal Variation of the Aberration Rates." *Mutat. Res.* 21:93-100.

Ma, T.-H., A. H. Sparrow, L. A. Schairer, and A. F. Nauman. 1978. "Effect of 1,2-dibromoethane (DBE) on Meiotic Chromosomes of *Tradescantia.*" *Mutat. Res.* 58:251-258.

Marshak, A. 1937. "The Effects of X-rays on Chromosomes in Mitosis." *Proc. Natl. Acad. Sci. U.S.A.* 23:362-369.

Matagne, R. 1968. "Chromosomal Aberrations Induced by Dialkylating Agents in *Allium cepa* Root-Tips and Their Relation to the Mitotic Cycle and DNA Synthesis." *Radiat. Bot.* 8:489-497.

Michaelis, A., and R. Rieger. 1971. "New Karyotypes of *Vicia faba* L." *Chromosoma* 35:1-8.

Mohandas, T., and W. F. Grant. 1972. "Cytogenetic Effects of 2,4-D and Amitrole in Relation to Nuclear Volume and DNA Content in Some Higher Plants." *Can. J. Genet. Cytol.* 14:773-783.

Nichols, W. W. 1973. "Significance of Various Type Chromosome Aberrations for Man." *Environ. Health Perspect.* 6:179-183.

Nicoloff, H., and K. Gecheff. 1976. "Methods of Scoring Induced Chromosome Structural Changes in Barley." *Mutat. Res.* 34:233-244.

Nicoloff, H., and G. Künzel. 1976. "Reconstructed Barley Karyotypes as a Tool in the Analysis of Intrachromosomal Distribution of Chromatid Aberrations." In *Barley Genetics*, vol. 3, ed. H. Gaul. Munich: Thiemig.

Nilan, R. A., and B. K. Vig. 1976. "Plant Test Systems for Detection of Chemical Mutagens." In *Chemical Mutagens*, vol. 4, ed. A. Hollaender, pp. 143-170. New York: Plenum.

Novick, A., and A. H. Sparrow. 1949. "The Effects of Nitrogen Mustard on Mitosis in Onion Root Tips." *J. Hered.* 40:13-17.

Perry, P., and H. J. Evans. 1975. "Cytological Mutagen-Carcinogen Exposure by Sister Chromatid Exchange." *Nature* (London) 258:121-125.

Perry, P., and S. Wolff. 1974. "New Giemsa Method for the Differential Staining of Sister Chromatids." *Nature* (London) 251:156-158.

Read, J. 1959. *Radiation Biology of* Vicia faba *in Relation to the General Problem.* Oxford: Blackwell.

Rushton, P. S. 1969. "The Effects of 5-fluorodeoxyuridine on Radiation-Induced Chromatid Aberrations in *Tradescantia* Microspores." *Radiat. Res.* 38:404-413.

Scheid, W. 1976. "Mechanism of Differential Staining of BUdR-Substituted *Vicia faba* Chromosomes." *Exp. Cell Res.* 101:55-58.

Schubert, I., G. Kuenzel, H. Bretschneider, and R. Rieger. 1980. "Sister Chromatid Exchanges in Barley." *Theoret. Appl. Genet.* 56:1-4.

Schubert, I., S. Sturelid, P. Döbel, and R. Rieger. 1979a. "Intra-chromosomal Distribution Patterns of Mutagen-Induced SCEs and Chromatid Aberrations in Reconstructed Karyotypes of *Vicia faba.*" *Mutat. Res.* 59: 27-38.

Schubert, I., P. Döbel, and S. Sturelid. 1979b. "Isostaining and Iso-nonstaining in 5-bromodeoxyuridine-Substituted *Vicia faba* Chromosomes." *Experientia* 35:592-593.

Schvartzman, J. B. 1979. "Three-Way Differentiation of Sister Chromatids in 5-bromodeoxyuridine-Substituted Chromosomes." *J. Hered.* 70: 423-424.

Schvartzman, J. B., and F. Cortés. 1977. "Sister Chromatid Exchanges in *Allium cepa.*" *Chromosoma* 62:119-131.

Schvartzman, J. B., and P. Hernandez. 1980. "Sister Chromatid Exchanges and Chromosomal Aberrations in 5-aminouracil-Synchronized Cells." *Theor. Appl. Genet.* 57:221-224.

Schvartzman, J. B., F. Cortés, A. Gonzalez-Fernandez, C. Gutiérrez, and J. F. Lopez-Saez. 1979a. "On the Nature of Sister-Chromatid Exchanges in

5-bromodeoxyuridine-Substituted Chromosomes." *Genetics* 92:1251-1264.

Schvartzman, J. B., R. Postigo, and C. Gutiérrez. 1979b. "Analysis of Visible Light-Induced Sister Chromatid Exchanges in 5-bromodeoxyuridine-Substituted Chromosomes." *Chromosoma* 74:317-328.

Sharma, A. K., and A. Sharma. 1972. *Chromosome Techniques: Theory and Practice.* London: Butterworths. Baltimore: University Park Press.

Shaw, M. W. 1970. "Human Chromosome Damage by Chemical Agents." *Ann. Rev. Med.* 21:409-432.

Shelby, M. D., and the Environmental Mutagen Information Center Staff. 1976. *Chemical Mutagenesis in Plants and Mutagenicity of Plant Related Compounds.* ORNL/EMIC-7. Oak Ridge, Tenn.: Oak Ridge National Laboratory.

Smith, H. H., and T. A. Lotfy. 1954. "Comparative Effects of Certain Chemicals on *Tradescantia* Chromosomes as Observed at Pollen Tube Mitosis. *Am. J. Bot.* 41:589-593.

Stich, H. F., and R. H. C. San. 1980. "International Workshop on Short-Term Tests for Chemical Carcinogens." *Genet. Soc. Can. Bull.* 11:25-28.

Strickberger, M. W. 1976. *Genetics.* 2nd ed. New York: Macmillan.

Swanson, C. P. 1957. *Cytology and Cytogenetics.* Englewood Cliffs, N.J.: Prentice-Hall.

Taylor, J. H., P. S. Woods, and W. L. Hughes. 1957. "The Organization and Duplication of Chromosomes as Revealed by Autoradiographic Studies Using Tritium-Labelled Thymidine." *Proc. Natl. Acad. Sci. U.S.A.* 43:122-128.

Tease, C., and G. H. Jones. 1979. "Analysis of Exchanges in Differentially Stained Meiotic Chromosomes of *Locusta migratoria* After BRdU-Substitution and FPG Staining. II. Sister Chromatid Exchanges." *Chromosoma* 73:75-84.

Van't Hof, J. 1965. "Discrepancies in Mitotic Cycle Time When Measured with Tritiated Thymidine and Colchicine." *Exp. Cell Res.* 37:292-299.

Vosa, C. G. 1976. "Sister Chromatid Exchange Bias in *Vicia faba* Chromosomes." In *Current Chromosome Research*, ed K. Jones and P. E. Brandham, pp. 105-114. Amsterdam: Elsevier/North Holland.

Wolff, S. 1979. "Sister Chromatid Exchange: The Most Sensitive Mammalian System for Determining the Effects of Mutagenic Carcinogens." In *Genetic Damage in Man Caused by Environmental Agents*, ed. K. Berg, pp. 229-246. New York: Academic Press.

——. 1963. *Radiation-Induced Chromosome Aberrations.* New York: Columbia University Press.

Wuu, K. D., and W. F. Grant. 1966a. "Morphological and Somatic Chromosomal Aberrations Induced by Pesticides in Barley (*Hordeum vulgare*)." *Can. J. Genet. Cytol.* 8:481-501.

——. 1966b. "Induced Abnormal Meiotic Behavior in a Barley Plant (*Hordeum vulgare* L.) with the Herbicide Lorox." *Phyton* 23:63-67.

2

Somatic Crossing-Over in Higher Plants

Baldev K. Vig

INTRODUCTION

In 1925 the Soviet geneticist Serebrovsky suggested the occurrence of somatic segregation in domestic fowl. This novel idea was based on his observation of apparently homozygous patches of feathers on the otherwise heterozygous background. Serebrovsky's suggestion brought to attention a parallelism between a fundamental meiotic phenomenon and an unexpected, rare activity of mitotic cells. Stern (1936), however, is generally credited with establishing a clear relationship between somatic crossing-over and segregation through his detailed and classical paper on *Drosophila*. By utilizing singed bristles (sn) and yellow body color (y) linked markers on heterozygous (sn y^+/sn$^+$y) flies, he showed the existence of homozygous recessive combinations for either gene in the form of singed or yellow single spots (sn y/sn y^+ and sn$^+$ y/sn y) as well as singed-yellow (sn y/ sn y) double spots.

The first serious suggestion of somatic recombination in a plant was found in the segregation for genes controlling color in *Zea mays* (Jones, 1937). It is not definitely known if this case represents somatic crossing-over, since variegation observed in the triploid endosperm can be explained by other mechanisms despite the fact that in these studies (Jones, 1937; 1938) even double spots were observed with high frequency (0.039 spots/kernel). The secondary twinning of red-white sectors on the background of dark sectors was also noticed.

Since these historical studies a host of organisms have been reported as suitable material for analysis of somatic crossing-over. These include many plant species, mammals, several insects, and fungi. Such studies have also brought into line the observations on mosaicism made prior to the time the concept of somatic crossing-over was established. Though a greater body of information on somatic crossing-over is available in the literature dealing with fungi, *Drosophila*, and the house fly, the present chapter confines its efforts to studies that deal primarily with higher plants. Also, even though terms like "somatic recombination," "somatic crossing-over," "somatic segregation," "mitotic segregation," "somatic variegation," and "mitotic crossing-over" have been used to define the observations of the type under discussion, the term "mitotic crossing-over" appears to be appropriate and precise. However, because of its historic significance, and the lack of precise evidence for occurrence of this phenomenon in mitotic cells of germinal tissues, I will use the term "somatic crossing-over" in this chapter. This term, therefore, may be considered equivalent to or interchangeable with other terms carrying similar connotation.

PLANT SYSTEMS AVAILABLE FOR DETECTION OF SOMATIC CROSSING-OVER

The most suitable systems for the detection of somatic crossing-over are the heterozygous diploid organisms or organs. In such systems the occurrence of this event results in two contiguous tissues with homozygous genotypes of each kind; for example, in a cell of a leaf tissue of the genotype *Aa*, somatic exchange between the centromere and the marker gene may produce *AA* and *aa* homozygous progeny. Depending upon the phenotype controlled by these genes, after several mitotic divisions the *aa* cells would appear as a sector or spot different from the *Aa* background. If *A* is incompletely dominant over *a*, the *AA*;*aa* spots would appear as a twin spot composed of two mirror images of similar size (see Figure 2.1).

Such twin spots were first reported by Dahlgren and Ossian (1927) in the apple epidermis. However, it was Jones (1937) who interpreted them, along with additional examples from citrus fruits, as having originated from somatic crossing-over. He also emphasized the notoriously heterozygous condition of vegetatively propagated fruits.

Since then several plant species have been reported to exhibit twin spots thought to have their origin through somatic crossing-over. A representative list of some of these species is given in Table 2.1 However, follow-up work has been carried out only with a few of these species, mostly dicotolydonous. The reason for the use of dicots is obviously the suitability of the leaf tissue, on which mutational events are expressed much more easily than on monocots. The developmental patterns in dicots are more favorable for the expression of genetic changes due to a large number of target cells that can divide and expand without much competition from the surrounding tissue. Additionally, even though sepals, petals, and seeds are favorable materials for study of sectors, such studies have been primarily limited to *Zea* and *Antirrhinum*.

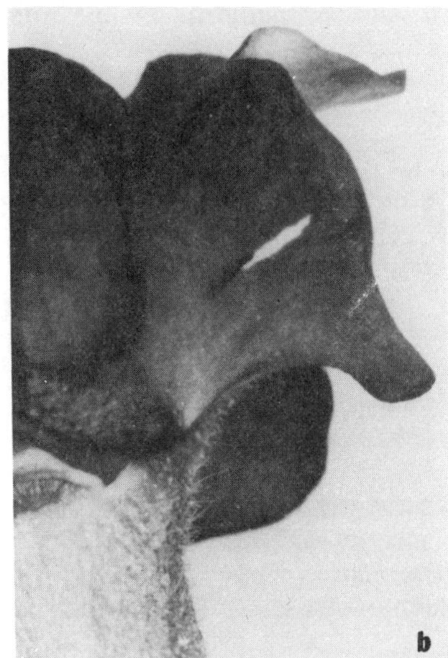

FIGURE 2.1. Examples of Twin Spots in Plants. (a) Spots on a simple leaf of a heterozygous $Y_{11}y_{11}$ plant of variety T219 of the soybean (*Glycine max*). The dark area is inferred to represent the $Y_{11}Y_{11}$ genotype and the light area, the $y_{11}y_{11}$ genotype as a result of somatic crossing-over in one of the ancestral cells. The leaf was in the primordial stage in the seed, which was treated with mitomycin C. (b) A twin spot on a petal of *incolorata/Incolorata* heterozygous snapdragon (*Antirrhinum majus*). The *inc/Inc* background is medium magenta. The *inc/inc* sector is acyanic (light color), whereas the *Inc/Inc* sector of the twin spot is dark magenta (black shade). Courtesy of Brian Harrison, John Innes Institute.

Glycine max

The soybean system is here singled out as an example of detailed study. A large number of chemicals have been used to induce and detect mosaicism in this organism. The occurrence of somatic crossing-over and associated chromosomal processes (exchanges, deletions, and so on) is thought to be responsible for induction of leaf spots (Vig, 1978; Vig, Nilan, and Arenaz, 1976; Vig, 1975b). In this case heterozygous $Y_{11}y_{11}$ plants (varieties T219, L65-1237, and L72-1937) have light green leaves, $Y_{11}Y_{11}$ plants have dark green

TABLE 2.1

Higher Plant Systems Promising for the Study of Mutagen-Induced Mosaicism, Including Twin Spots

Species	*Tissue*	*Genes Involved*	*Spontaneous Spot Frequency*	*Reference*
Antirrhinum majus	petal	*niv, inc, pal, eos*		Harrison and Carpenter (1977)
Arabidopsis thaliana	leaf	ch^1 (*pa*)		Ahnström, Natarajan, and Veleminsky (1972) Hirono and Rédei (1965)
Collinsia heterophylla	leaf	w_1		Zaman and Rai (1973)
Glycine max	leaf	y_{11}	0.08/leaf	Evans and Paddock (1979)
Gossypium barbadense	leaf	v_1, v_7	0.33/leaf	Barrow, Chaudhari, and Dunford (1973)
Hordeum vulgare	leaf			Nilan (unpublished)
Lycopersicon esculentum	leaf	Xa_2	0.01/leaf	Ross and Holm (1959)
Nicotiana tabacum	leaf	a_1, a_2, *t, su*	0.15/leaf (for *su*)	Deshayes and Dulieu (1974) Carlson (1974)
Petunia hybrida	leaf, petal	*a, b, p, r*	$\simeq 0$	Cornu (1970); Cornu and Dommergues (1974)
Salvia splendens	petal	*v, l, int, p*	0.3/petal	Hendrychova-Tomkova (1964)
Tradescantia hirsuticaulis	stamen hair, petal	*d, e*	0.001/stamen hair	Cuany, Sparrow, and Jahn (1959); Sparrow, Schairer, and Villalobos-Pietrini (1973); Mericle and Mericle (1974); Christianson (1975); Nauman, Sparrow, and Schairer (1976)
Zea mays	endosperm, leaf	*c, pr, su, wx, yg*	0.039/kernel	Jones (1937); Conger and Carabia (1977)

Source: Modified from Evans and Paddock (1979, Table 2), with permission.

leaves, and $y_{11}y_{11}$ plants have golden yellow leaves. Whereas the leaves of plants obtained from fresh seed are usually unremarkable, those obtained from older seed generally show spots on the two simple leaves and, occasionally, on the first compound leaf. The frequency of spots is dependent upon the age of the seed, the storage conditions, and growth conditions.

The spots on the $Y_{11}y_{11}$ leaves are dark green (resembling $Y_{11}Y_{11}$ leaves), yellow (like $y_{11}y_{11}$ leaves), or twin (double). The double spots are composed of a dark green spot adjacent to, and a mirror image of, a yellow spot. These double spots are inferred to originate through somatic crossing-over that leads to the formation of $Y_{11}Y_{11}$-$y_{11}y_{11}$ sectors on the $Y_{11}y_{11}$ background. Some of the single (dark green or yellow) spots may have their origin in failure of one of the $Y_{11}Y_{11}$-$y_{11}y_{11}$ components forming the initial twin (complementary) cells; others may arise from chromatid segmental exchanges, losses, gains, or numerical inequalities during mitosis (Vig, 1969), or from simple point mutations (Vig, 1973a; 1973b; Nilan and Vig, 1976). The light green spots (like $Y_{11}y_{11}$ leaves) on the $y_{11}y_{11}$ leaves can best be explained by the occurrence of point mutations. Some light green and very dark green spots, however rare, are also seen on $Y_{11}Y_{11}$ leaves; these may suggest formation of $Y_{11}0$-$Y_{11}Y_{11}Y_{11}$ cells. However, only the twin spots can be taken as evidence of somatic crossing-over having occurred.

Nicotiana tabacum

In *Nicotiana tabacum* several genes (such as a_1, a_2, t, su) have been utilized to determine the occurrence of somatic crossing-over. Among the higher plants this is the only organism in which linked markers have been used to account for the occurrence of twin spots through reciprocal recombination in mitotic cells. Carlson (1974) used the genes t and su as linked markers to produce twin spots of the constitution $t^+su/t\ su$ and $t\ su^+/t\ su$ on the $t^+\ su/t\ su^+$ background. He regenerated whole plants through tissue culture of spots to elucidate the genetic constitution of the spots. Deshayes and Dulieu (1974) and Dulieu (1975) recovered a_1/a_1 homozygotes by regeneration of mesophyll explants from the spots on a_1^+/a_1 leaves. Whereas in Carlson's studies 11 of 12 spots were inferred to be due to true somatic crossing-over (1 of 12 being the result of nondisjunction), the data in Dulieu's studies proved the occurrence of reciprocal exchanges between homologues as well as nonhomologues, in addition to deletions, gene conversions, and point mutations. These studies have given credence to the concept of the occurrence of somatic crossing-over in higher plants more than any other data available.

Arabidopsis thaliana

The plants heterozygous for the gene ch^1, *pa* used in studies by Hirono and Rédei (1965) and Ahnström, Natarajan, and Veleminsky (1972) display sectors apparently homozygous for the gene ch^1/ch^1. Even though critical analysis of the sort possible in *Nicotiana* has not been carried out, the occurrence of spots can be attributed to somatic or premeiotic recombination (Hirono and Rédei, 1965) or to a phenomenon "like somatic recombination" (Ahnström, Natarajan, and Veleminsky, 1972). Also, since *Arabidopsis thaliana*, unlike *Glycine* or *Nicotiana*, is a true diploid, any deletion of substantial genetic material is likely to lead to death of the affected cells. The spots of apparently ch^1/ch^1 constitution are so large (sometimes covering one-third to one-half of the leaf) that only cells with a complete genome are likely to survive and reproduce without being competed out by the surrounding diploid, heterozygous tissue.

Tradescantia hirsuticaulis

Even though early studies by Sparrow (see Cuany, Sparrow, and Jahn, 1959; Sparrow, Schairer, and Villalobos-Pietrini, 1973, for example) pointed to mechanisms like gene mutations as inducing single as well as twin spots on the stamen hairs in stocks heterozygous for genes *d* and *e*, later studies by Mericle and Mericle (1974) and by Christianson (1975) strongly indicate somatic crossing-over as the origin of twin spots on stamen hair and petals. In this case the incidence of twinning can be dramatically increased by the application of mitotic recombinogens.

Antirrhinum majus

The snapdragon offers an illustration of twin sectors on the petals of *Salvia splendens* heterozygous for genes, *v*, *l*, *int*, and *p* been used superbly to demonstrate the occurrence of twin spots due to somatic crossing-over and also to point out the biochemical constitution of the spots by precursor analysis (Harrison and Carpenter, 1973; 1977; Harrison and Stickland, 1974). In a study using inc^+ $eos/inc\ eos^+$ plants, $inc\ eos^+/inc\ eos^+$ sectors were produced, both of which differed from the medium magenta background in being acyanic and dark magenta in color, respectively.

Other Species

Double spots have been observed on leaves of *Xa-2/xa-2* plants of *Lycopersicon esculentum* (Ross and Holm, 1959) and on the petals of *Salvia splendens* heterozygous for genes *v*, *l*, *int*, and *p* (Hendrychova-Tomkova, 1964). The genes *p* and *v* are located on the same chromosome of *Salvia splendens*, yet their very close linkage (1.5 percent recombination) has been of no help in definitely establishing the occurrence of somatic crossing-over.

In *Petunia hybrida* several genes (*a*, *b*, *p*, *r*) are suitable for the study of twin spots. However, only limited data can be taken as an expression of somatic crossing-over. One such possibility has been expressed for the gene *a* (Cornu, 1970), even though detailed experiments are possible with heterozygous a^+r/ar^+ genotypes in which deletions, point mutations, and gene conversions have been inferred to be the causes of single spots (Cornu and Dommergues, 1974).

Twin spots are also found in *Gossypium barbadense* heterozygous for genes v_7 and v_1. Such spots have correctly been assigned to the process of homeologous translocation (Barrow and Dunford, 1974), since even v_1^+/v_1 and v_7/v_7^+ tetraploid, as well as v_1^+, v_7 diploid, plants show twin spots (Barrow, Chaudhari, and Dunford, 1973).

A claim by Zaman and Rai (1973) that single spots on the leaves of w_1^+/w_1 *Collinsia heterophylla* are due to somatic crossing-over can be disputed, since no evidence like that for production of twin spots is available.

In summary, whereas several plant species express twin spots on the heterozygous background (see Table 2.1), and the components of their twin spots appear to have homozygous genetic makeup, critical studies using linked markers or regeneration of mesophyll tissue are necessary before other mechanisms can be ruled out, leaving somatic crossing-over as the only mechanism. Circumstantial evidence, however, may rule out all mechanisms except somatic crossing-over as responsible for the production of twin spots.

MECHANISMS POSSIBLE AS ORIGIN OF SOMATIC CROSSING-OVER

There is little direct evidence available for any possible mechanism involved in the origin of somatic crossing-over in higher plants. The circumstantial evidence to support the occurrence of some mechanisms reminiscent of those involved in meiotic crossing-over comes from somatic pairing observed in root tips of *Vicia faba* (Kitani, 1963), and possible somatic association (spatial closeness) of homologous chromosomes in wheat (Avivi and Feldman, 1973; Avivi, Feldman, and Bushuk, 1969), *Impatiens*, and *Salvia* (Chauhan and Abel, 1968). In this regard the plant chromosomes in somatic cells appear to have behavior similar to that of animal chromosomes, such as in the Tasmainian rat kangaroo (Gibson, 1970) or humans (German, 1964). The origin or direct proof of significance of such associations is, however, lacking. Such associations may possibly result from pairing of heterochromatic centromeric regions of homologous chromosomes in the interphase cell (see Vig, 1975a). If so, one should find that the occurrence of somatic crossing-over more often involves genes in the paracentromeric regions. Such is the case in *Schizosaccharomyces pombe*, in which a study of the meiotic/mitotic recombination ratio clearly points to a far higher incidence of somatic crossing-over for the genes close to the centromere (Minet, Grogsenbacher-Brunder, and Thuriaux, 1980). In both plants and animals, meiotic recombination is known to be rare in this region of the chromosome.

Critical cytological evidence depends upon correlating the formation of a chiasmalike configuration in mitotic cells of a tissue with the frequency of occurrence of mosaic sectors in that tissue. Such correlations are entirely lacking in plant materials. Evidence of formation of chiasmalike configurations in mitotic cells has been given by German and La Rock (1969) and Rao and Natarajan (1967) in cells treated with mitomycin C, which is an effective mitotic recombinogen. These chromosomes showed exchanges of equivalent genetic material at corresponding sites, detected cytologically. Their localization in heterochromatic regions (Shaw and Cohen, 1965) supports the idea of heterochromatic pairing in the mitotic interphase (Vig, 1975a). Huttner and Ruddle (1976) also provided evidence of mitomycin C-induced somatic chiasmata in chromosomes that had been made to incorporate BrdUrd for detection of such exchanges. Mitotic chiasmata have also been observed in a human disease, Bloom's syndrome, without any correlation with genetic crossing-over (German, 1964; Kuhn, 1976).

There are some reports to correlate somatic crossing-over with cytologically visible chiasmata in mitotic chromosomes. Merriam, Nothiger, and Garcia-Bellido (1972) provided evidence of the formation of dicentric anaphase bridges resulting from somatic crossing-over in X-chromosome inversion heterozygotes of *Drosophila melanogaster*. An elegant series of photographs by Rubini, Vecchi, and Franco (1980) show chiasmalike configuration in both sexes of *Musca domestica* at all stages of development. It provides rather convincing evidence of correlation between somatic crossing-over detectable phenotypically and cytological phenomena manifest in its occurrence.

FREQUENCY OF SPONTANEOUS SOMATIC CROSSING-OVER

A precise estimate of the frequency of spontaneous somatic crossing-over is an exercise in futility. First, the diagnostic measures of somatic crossing-over, twin spots, are a manifestation only of the successful phenotypic expression of original events. It is not known what fraction of the original events takes place at sites suitable for development of the spot and what fraction of the total spots is macroscopically visible. Second, the physical conditions of growth chambers or the greenhouse, such as temperature, are known to affect the frequency of visible twin spots. Third, it is possible that certain strains of a given plant species exhibit higher or lower frequency of somatic recombination than that found in normal genotypes. Fourth, even in the same organism, different genes will undergo somatic crossing-over differently because of their location on the chromosome, nearness to heterochromatic block, and the presence of "hot spots" of breakage/reunion at suitable sites along the chromosome. For example, in *D. melanogaster*, X-ray induced somatic crossing-over is far more frequent for the genes located on the X chromosome than for those on chromosome 3 (Becker, 1956), perhaps because of a greater amount of heterochromatin on the X chromosome. Last, all conditions being equal, the developmental stage of the organism has a strong bearing on the frequency of induced somatic crossing-over; for example, in *Drosophila* older larvae show a reduced effect of radiation on this phenomenon (Abbadessa and Burdick, 1963). After all, even spontaneous somatic crossing-over is induced by either physical or chemical agents. Also, intragenic recombination is expected to differ from intergenic recombination. It is of interest that whereas *D. melanogaster* males and females both show spontaneous somatic crossing-over (as in Stern, 1936), *M. domestica* expresses it only with exogenous treatments like radiations (Nothiger and Dubendorfer, 1971).

The impact of the above-mentioned factors on the frequency of visible sectors is compounded when one considers interspecific or intergeneric comparisons. In the yeast *Saccharomyces cerevisiae*, one can estimate the number of treatable cells and the products of crossing-over rather accurately. On the other hand, in higher organisms or their tissues the number of cells and the layers of cells involved in such a process are only very rough estimates.

Even then, the observed frequency of somatic crossing-over—or twin spots, to be more precise—is roughly comparable for organisms widely diverse on the evolutionary scale. As shown in Table 2.2, the frequency in higher plants differs by no more than one order of magnitude between species or genera, including fungi, where it ranges between 1 to 12×10^{-5} events/cell (see Evans and Paddock, 1979). One may generalize that the frequency is in the neighborhood of 10^{-5} to 10^{-4} events/cell.

INDUCTION OF SOMATIC CROSSING-OVER IN HIGHER PLANTS

Direct-Acting Chemicals

Mitomycin C

Several chemicals are known to be capable of increasing the frequency of somatic recombination. This is also true of radiations

TABLE 2.2
Comparative Rates of Spontaneous Frequencies of Somatic Crossing-over in Various Organisms

Organism	*Frequency/cell*
Antirrhinum majus	0.96×10^{-5}
Nicotiana tabacum (in vitro)	4.6×10^{-5}
Nicotiana tabacum (in vitro)	6.7×10^{-5}
Nicotiana tabacum (field)	0.77×10^{-5}
Glycine max	12.8×10^{-5}
Tradescantia hirsuticaulis	13.5×10^{-5}
Musca domestica	0

Source: Vig (1978), modified from Evans and Paddock (1979, Table 13), with permission.

to a certain degree. However, the most effective agent is mitomycin C. Even though it is like dialkylating agents in its action on the genetic material (Iyer and Szybalsky, 1963; Vig, 1978), it differs from such agents, and all others, in being extremely effective in inducing crossing-over in mitotic cells. The induction of somatic crossing-over by mitomycin C is discussed in some detail below, since it is considered a standard mitotic recombinogen.

The first report of an increase induced in the incidence of somatic crossing-over by mitomycin C was in the fungi *Ustilago maydis* and *S. cerevisiae* (Holliday, 1964). At concentrations of 1.2 mM in the agar medium, Holliday exposed these fungi for 200 minutes and obtained a 32-fold and eight fold increase in the expression of the phenomenon in *Ustilago* and *Saccharomyces*, respectively. Besides, unlike ultraviolet, mitomycin C at these subfungicidal doses did not increase the mutation frequency in haploid *Ustilago*.

The incidence of somatic crossing-over has been increased by the use of mitomycin C in *Glycine max* (Vig and Paddock, 1968) and *Nicotiana tabacum* (Carlson, 1974). It is considered instructive to reproduce some data with one of these systems, *Glycine max*.

The early experiments showed the capability of mitomycin C as a somatic recombinogen in *Glycine*. Concentrations as low as 0.000325 percent for 24 hours up to as high as 0.005 percent for 24 hours induced a significant increase in the frequency of the three types of spots on $Y_{11}y_{11}$ leaves. The highest relative increase was that for double spots, as if it were the major effect of mitomycin C. Thus, the chemical brings about complementary segregation for the gene pair $Y_{11}y_{11}$ through the induction of somatic crossing-over as its primary effect. The failure of either component of the products of somatic crossing-over would produce a single spot—dark green or yellow—and the two types should be nearly equal in frequency. This is generally true, and leads one to believe that single spots originate from such failures or from chromosome aberrations like exchanges between nonhomologues, deletions, and so on. A summary of representative results is given in Table 2.3.

In *Glycine* the two simple leaves have a far higher frequency of spots than do the compound leaves. This reflects the number of initial embryonic cells present in the seed. The absence of any spots on second and subsequent compound leaves in plants treated as seed with mitomycin C points to the fact that the chemical has no systemic effect.

That the chemical has no systemic effect was further tested by application of 0.005 percent mitomycin C in lanolin paste to the third and fourth compound leaves after the first compound leaf had

TABLE 2.3
Spots on the Two Simple Leaves and the First Compound Leaf of Soybean Var. T219 Obtained from Seed Treated with Mitomycin C

Treatment		No. Leaves Analyzed	*Spot Frequency per Leaf*			X Control
			DG	*Yl*	*Db*	
A.						
Control		230	0.35	0.15	0.06	
Mitomycin C						
0.000325%	24 hrs.	150	0.75	0.50	0.65	10.8
0.00065%	24 hrs.	130	1.25	1.39	0.89	14.8
0.00125%	24 hrs.	185	1.62	1.40	1.79	29.8
0.0025%	24 hrs.	175	1.15	1.23	1.90	31.6
B.						
Control		125	0.20	0.02	0.04	
Mitomycin C						
0.000625%	24 hrs.	175	1.01	0.46	1.61	40.2
0.00125%	24 hrs.	105	2.44	1.09	2.17	54.2
0.0025%	24 hrs.	125	1.93	0.84	2.37	59.2
C.						
Control		90	0.02	0.19	0.06	
Mitomycin C						
0.0025%	12 hrs.	115	2.07	1.91	1.72	28.7
0.0025%	24 hrs.	110	4.43	3.85	6.65	110.8
0.005%	12 hrs.	145	3.59	2.72	3.04	50.7
0.005%	24 hrs.	110	3.82	3.09	3.12	52.0

Note: DG = dark green; Yl = yellow; Db = double or twin; X control = treated/control value for doubles only.

Sources: A and B, Vig and Paddock (1968); C, Vig (1973a).

partially unfolded. A frequency, in one experiment, of 0.27 double spots/leaf, compared with the frequency of 0.005 in the control (Vig and Paddock, 1968), proved the point.

The effect of mitomycin C varies with the physiological age of the seed (Vig, 1973c). Seeds were treated with aqueous solution (0.0025 percent) during 0-36 hours for a period of only 4 hours—0-4 hours, 4-8 hours, 8-12 hours, and so on. A cyclic pattern of increase in the frequency of spots emerged. The frequency of spots increased in treatments lasting 16 hours after germination. Then a decline was observed, followed by another sharp increase at about 24-28 hours. This was followed by another drop. The results from one set of such a study are given in Table 2.4.

These data indicate a sort of correlation between the frequency

TABLE 2.4
Frequency of Different Types of Spots on the Leaves of Soybean Var. L65-1237 Treated with 0.0025% Mitomycin C during Different Periods of Germination

Treatment Period (Hrs.)	*Number of Leaves Analyzed*	*Spot Frequency and Type per Leaf*		
		DG^+	*Yl*	*Db*
1. —	135	0.33	0.18	0.22
2. 8 (H_2O)	85	0.19	0.29	0.19
3. 0-4	80	0.35	0.36	0.33
4. 4-8	110	0.56	0.49	0.83
5. 8-12	140	0.67	0.95	1.41
6. 12-16	145	1.03	0.95	1.34
7. 16-20	105	0.64	0.50	0.78
8. 20-24	140	1.51	1.51	1.72
9. 24-28	150	0.73	0.92	0.95
10. 28-32	115	0.78	1.35	1.30
11. 32-36	140	1.01	1.56	1.26

Note: DG^+ = dark green; Yl = yellow; Db = double or twin. 0 = time of initiation of soaking the dry seeds.

Source: Vig (1973c).

of induced spots and the synthetic activity of the DNA molecule. DNA synthesis, as measured by ^{3}H-TdR incorporation, appears to start in soybean embryos at about 24 hours after initiation of germination, and increases until about 42 hours. This increase stays on a plateau for about 20 hours, followed by another burst of activity (^{3}H-TdR incorporation) until at least the eightieth hour (Vig, 1973c). It is possible that an increase in the frequency of spots depends upon the duplication of the chromosome segment carrying the gene Y_{11} and/or that DNA synthesis is required for completing reciprocal rejoining.

In addition, the data did not present a definite correlation with uptake of ^{3}H-arginine as a cause of induction of spots (Vig, 1973c), even though the cross-linking efficiency of mitomycin C is dependent upon the presence or absence of proteins associated with DNA (Szybalsky, 1965).

One expects the action of mitomycin C to be dependent upon the concentration of ATP in the cell, since somatic crossing-over involves repair. Sodium azide, an inhibitor of oxidative phosphorylation that interferes with chromosome repair, was used in combination with mitomycin C (Vig, 1973b). It does not cause chromosome breakage per se (Sander et al., 1978). The idea was that an inhibition of somatic crossing-over by NaN_3 would create free fragments and, hence, single spots, if the phenomenon utilized a breakage rejoining sequence. However, if complementary exchanges are brought about by cross-linking or an alkylating property of the mitomycin C molecule, the NaN_3-induced changes may not interfere in the process required for DNA synthesis.

Several experiments were carried out by treating the seed with 0.01 or 0.02 percent aqueous solution of NaN_3 with or without 0.005 and 0.001 percent mitomycin C. NaN_3 alone increased the frequency of dark green and yellow spots sixfold to ninefold, but only twofold to threefold for the doubles (Vig, 1973b). Mitomycin C increased the frequencies of all three types of spots equally, approximately twofold to 2.5-fold for 0.0005 percent treatment and fourfold to fivefold for 0.001 percent treatment. The effect of the two chemicals was merely additive, and no synergism was operative. The data permit the conclusion that mitomycin C does not act synergistically with NaN_3 in producing spots in *Glycine*.

The incidence of somatic crossing-over is reduced by application of a nucleoside, particularly deoxyribose cytidine (Vig, 1972). Similar effects have been reported by Ashley (1978) when colchicine treatment is given simultaneously with mitomycin C. These studies may suggest that application of colchicine or nucleosides interferes with interphase chromosome arrangement and with pairing of corresponding segments of the heterochromatic zones of homologous chromosomes (Vig, 1975a). However, pretreatment with colchicine has been reported to produce a mild synergistic effect with mitomycin C (Vig, 1971).

These data using mitomycin C, therefore, permit the conclusions that mitomycin C is capable of inducing somatic crossing-over in the *Glycine* system, that it shows additive effects with NaN_3 and a synergistic effect with colchicine, and that it is inhibited in its effect by deoxyribose cytidine.

As in the studies with *Glycine*, an increase in the frequency of twin spots has been induced in *Nicotiana tabacum* treated with mitomycin C. Carlson (1974) treated suspension cultures of tobacco with 0.1 mM of the chemical and observed a 4.7-fold increase over the spontaneous frequency of double spots.

Evans and Paddock (1979) subjected the data reported by Vig to regression analysis based on the increase of twin spots/μM mitomycin C applied to seed. They also generated some data with X rays, using similar material and experimental conditions (Evans, 1977). In both cases the rate of increase of double spots was linear for the treatment. Even though the regression coefficients for yellow and dark green spots in soybeans were similar for 1 μM mitomycin C versus 1 R X rays, for double spots the coefficient for mitomycin C was about six times the value for X-ray-induced double spots. These results indicate that "mitomycin C preferentially increases the frequency of genetic exchange via mitotic crossing-over" (Evans and Paddock, 1979, p. 330).

Caffeine

The induction of twin spots in *Glycine*, *Nicotiana*, and *Antirrhinum* also has been possible through application of an oxypurine, caffeine. In *Glycine* doses as low as 0.063 percent applied to seed for 4 hours in aqueous solution induce more than doubling of all types of spots. In one material treated for four hours, doubles increased sevenfold with 0.063 percent, 19-fold with 0.125 percent, 33-fold with 0.25 percent, and 71-fold with application of 0.5 percent caffeine (Vig, 1973a). Increase was also observed on the lower surface of the leaf, even though spots are rare on this surface. Carlson (1974) reported a 1.9-fold increase in the incidence of twin spots in *Nicotiana* callus given 0.5 mM caffeine in the medium.

Harrison and Carpenter (1977) induced twin spots in *Antirrhinum* with caffeine. They found that the frequency of twin spots is not affected by insertion of unstable genes *nivea-recurrens* and *pallida-recurrens* in the genetic background of the plant. The same genes increase the frequency of single spots. However, when a 0.25 percent or 0.5 percent solution of caffeine was supplied by immersing inflorescences for 24 hours a 32-fold or 123-fold increase was observed in the frequency of twin spots on the *Inc/inc* genotype. A 52.5-fold increase was observed on petals of the *Eos Niv/eos niv* genotype treated with 0.25 percent caffeine for 24 hours. Similarly, the *Niv Pal Sul/niv pal-tub sul* genotype responded with 52-fold and 40-fold increase in twin spots when 0.25 percent and 0.5 percent caffeine solutions (24 hours), respectively, were applied. Increases also were observed in the frequency of single spots.

Other Chemicals

A host of chemicals other than those mentioned above have been tried in various systems. These include inhibitors of DNA synthesis (such as 5-fluorodeoxyuridine), alkylating agents (ethyl methanesulfonate), nitrofurans (nifurprazinum), inhibitors of protein synthesis (puromycin), inhibitors of RNA synthesis (actinomycin D), inhibitors of oxidative phosphorylation (sodium azide), nitrosoureas and nitrosamines, and spindle depolymerizers (colchicine) (see Table 2.5). However, not all of these produce twin spots. As expected, those effective in producing double spots vary greatly in their efficiency, which, besides other factors, is a function of structural modification of the chemical. However, those effective in inducing double spots appear to have a common property: cross linking the DNA strands. The present paper does not provide space for detailed discussion of the possible mechanism(s) of action of these chemicals, and it is suggested that the reader find this information in the original papers referred to in Table 2.5.

TABLE 2.5
Mutagens Found Effective in Increasing the Frequency of Spots in Higher Plants

Mutagen	*Test System*	*Twin Spots Induced*	*Reference*
Actinomycin D	*Glycine max*	-	Vig (1973a)
Caffeine	*G. max*	+	Vig (1973a)
	Antirrhinum majus	+	Harrison and Carpenter (1977)
Colchicine	*G. max*	+	Vig (1971)
	Nicotiana tabacum	+	Carlson (1974)
	Lycopersicon esculentum	+	Ross and Holms (1959)
Chloramphenicol	*Arabidopsis thaliana*	-	Ahnström, Natarajan, and Veleminsky (1972)
1,2 dibromo-ethane	*Zea mays*	-	Nauman, Sparrow, and Schairer (1976)
Diepoxybutane	*G. max*	+	Vig and Zimmermann (1977)
Dimethyl-nitrosoamine	*G. max*	+	Arenaz and Vig (1978)
Ethyl methane-sulfonate	*G. max*	+	Vig, Nilan, and Arenaz (1976)
	Z. mays	-	Conger and Carabia (1977)
	Petunia hybrida	-	Cornu (1970); Cornu and Dommergues (1974)
5-fluourodeoxy-uridine	*G. max*	-	Vig (1973)
Furylfuramide	*G. max*	-	Fujii (1980)
Methyl butane-sulfonate	*G. max*	+	Vig, Nilan, and Arenaz (1976)
Methyl ethane-sulfonate	*G. max*	+	Vig, Nilan, and Arenaz (1976)
Methyl methane-sulfonate	*G. max*	+	Vig, Nilan, and Arenaz (1976)

(continued)

TABLE 2.5 (Continued)

Mutagen	*Test System*	*Twin Spots Induced*	*Reference*
Methyl nitroso-amine	*G. max*	+	Arenaz and Vig (1978)
Methyl nitrosourea	*A. thaliana*	-	Ahnström, Natarajan, and Veleminsky (1972)
Mitomycin C	*G. max*	+	Vig and Paddock (1968)
Nifurprazinum	*G. max*	+	Vig and Zimmermann (1977)
Psoralin	*N. tabacum*	+	Carlson (1974)
Puromycin	*G. max*	-	Vig (1973a)
Sludge (paper mill waste)	*G. max*	+	Klekowski and Levin (1979)
Sodium azide	*G. max*	-	Vig (1973b)
	Z. mays	-	Conger and Carabia (1977)
	A. thaliana	-	Ahnström, Natarajan, and Veleminsky (1972)
Trenimone	*G. max*	+	Vig and Zimmermann (1977)
γ rays	*P. hybrida*	+	Vig and Dulieu (unpub.)
	G. max	+	Vig (1974)
	T. hirsuti-caulis	+	Sparrow, Schairer, and Villalobos-Pietrini (1973); Christianson (1975); Nauman, Sparrow, and Schairer (1976); Cuany, Sparrow, and Jahn (1959); Mericle and Mericle (1974;1967)
	N. tabacum	+	Carlson (1974)
^3H_2O	*G. max*	+	Vig (1974)
Ultraviolet	*N. tabacum*	+	Carlson (1974)
X rays	*A. thaliana*	+	Hirono and Rédei (1965)
	P. hybrida	+	Cornu (1970); Cornu and Dommergues (1974)
	N. tabacum	+	Evans and Paddock (1979)
	G. max	+	Evans and Paddock (1979)

Source: Modified from Evans and Paddock (1979, Table 8), with permission.

Chemicals Requiring Metabolic Activation

Several promutagens have been converted to active mutagens with the help of the S9 fraction of the liver homogenate. Plant studies have indicated that the liver is not the only system that has enzymatic machinery capable of such activation. In 1968, Veleminsky and Gichner showed the effectiveness of some nitroso compounds in plants that require liver homogenates to be effective in animal tissue cultures or bacteria. However, their studies did not deal with somatic crossing-over.

In the soybean it has been possible to increase the frequency of twin spots by the use of the promutagen dimethyl nitrosamine (Arenaz and Vig, 1978). When seeds were treated with concentrations of the chemical as low as 1.25 ppm for 24 hours, a 2.8-fold increase in the frequency of all types of spots was found. There was a similar increase in the frequency of twin spots. At a concentration of 25 ppm, the increase in twin spots was about fivefold, and at 60 to 500 ppm, up to 39-fold. Interestingly, a maximum response to all types of spots was found at concentrations of 60 ppm. The data might mean that there is maximum conversion of the promutagen to mutagen at 60 ppm or less. This idea is supported by the observation that a related chemical, methyl nitrosourea, is effective in increasing the frequency of double and single spots up to a concentration of 125 ppm, which is tolerated without causing mutilation and growth retardation of the plant and leaves. In our studies dimethyl nitrosoamine could be tolerated up to 500 ppm, apparently because of its nonconversion to a toxic mutagen.

Radiations

The frequency of double spots has been increased in various species with ^3H_2O, X rays, and γ rays. The only example of ^3H-emitted β particles is that of *Glycine max*, in which concentration of ^3H as low as 5 $\mu C/ml$ were found effective on plants raised from seed exposed for 92 hours. A clear response was found in material treated with 50 $\mu C/ml$ for 38 or 62 hours, and 100 $\mu C/ml$ for 96 hours (Vig, 1974). Similarly, 50 R of γ radiation from a ^{60}Co source produced about a 1.7-fold increase in the frequency of doubles; with 100 R the increase was 2.1-fold, and with 250 R it was 6.7-fold. With a 1,000 R dose of X ray, Evans (1980) obtained a 12.47-fold increase in the frequency of double spots on the simple leaves of irradiated seeds with a moisture content of 55 percent and a 23.64-fold increase in seeds irradiated dry (6–8 percent moisture content). The increase of such spots on the first compound leaf was 5.43-fold for moist seeds and 1.81-fold for dry seeds. The interpretation that the moisture content of seed had no effect on the frequency of spots induced by X rays needs confirmation, since an unequivocal effect of moisture content on *Glycine* seed treated with γ rays has been shown (Vig, 1974). It must also be kept in mind that in all these experiments an increase in the frequency of single yellow or dark green spots on the $Y_{11}y_{11}$ background was also observed.

Twin spots have been induced with "about 13000 R of x-rays" in *Arabidopsis thaliana* (Hirono and Rédei, 1965), and a 3.1-fold increase was induced in *Nicotiana tabacum* with 900 R of ^{60}Co-emitted γ rays (Carlson, 1974). In some preliminary studies Vig and Dulieu (unpublished) induced twin sectors on the leaves of heterozygous *Petunia hybrida* treated with X radiation.

In the only report on ultraviolet, Carlson (1974) induced a 4.3-fold increase in the frequency of doubles in *Nicotiana* by irradiating callus tissue with 1,000 ergs/mm^2. The recombination obtained by X rays in *Lycopersicon esculentum*, however, appears to be lower in frequency than in other plant species (Evans, 1977). One may also find additional examples in *Tradescantia* (Mericle and Mericle, 1967), *Antirrhinum* (Cuany, Sparrow, and Jahn, 1959), *Zea* (Stein and Steffensen, 1959), *Arabidopsis* (Ahnström, Natarajan, and Veleminsky, 1972), and *Petunia* (Moore and Haskins, 1935). Also, in *Tradescantia hirsuticaulis* a 34 R exposure did not increase the frequency of red-blue twin sectors on stamen hair, whereas a 60 R exposure increased the frequency about 1.5-fold (Christianson, 1975). These sectors were found not to be associated with chromosome fragments scored as micronuclei.

Temperature

It is well known that the frequency of meiotic crossing-over increases at higher temperature, whereas that of somatic crossing-over decreases. Thus, the increased induction of somatic crossing-over by lowering the temperature can be used to test for the occurrence of this phenomenon. Such observations have been made in *D. melanogaster* (Brosseau, 1957). In plants an example is *Glycine*, in which plants raised at 37°C for a period of six hours every day (maintained at 74°F day/64°F night) exhibited more twin spots than did contols (74°/64°F) (Vig and Paddock, 1970). Such experiments need to be carried out with *Nicotiana* and other plant species amenable to the study of somatic crossing-over in order to establish the universality of the phenomenon.

EVIDENCE FOR OCCURRENCE OF SOMATIC CROSSING-OVER

Direct evidence for the occurrence of somatic crossing-over in plants is scant, and limited to *Arabidopsis thaliana* and *Nicotiana tabacum.* In the former, the linked markers *gi-ch-pa* were used in *gi + pa/ + ch +* combination by Hirono and Rédei (1965). Besides observing twin sectors (yellow-green and dark green), they subjected some sectors from which shoots had appeared to progeny test. The data showing some exchange of linked markers were believed to represent "an induced exchange at a four-strand stage in the meristematic cells." The occurrence of somatic pairing of chromosomes in *Arabidopsis* makes the suggestion even more plausible.

In 1971, Dulieu, de Boelpaepe, and Deshayes reported the occurrence of somatic recombination between genes a_1 and a_2 in *Nicotiana xanthi.* Later (Dulieu, 1972) a technique for the isolation and regeneration of sectors from leaves of *Nicotiana tabacum* was reported. Carlson (1974), combining the two ideas, reported the occurrence of reciprocal somatic exchanges for the genes *cl* and *su.* Using these linked markers, he clearly demonstrated the occurrence of somatic crossing-over in tobacco suspension cultures treated with a variety of chemical and physical agents. On the basis of chromosome counts in regenerated plants, he observed that 11 of 12 recombinants originated from somatic crossing-over—one was the result of nondisjunction.

The occurrence of acyanic/dark magenta, acyanic/pale, and pale/dark twin spots on *Eos:Niv/eos:niv* plants and acyanic/pale magenta, acyanic/dark, pale/dark, and pale/medium twin spots in *Niv:Pal:Sul/niv:pal:sul* genotypes of *Antirrhinum majus* (Harrison and Carpenter, 1977) strongly suggests somatic crossing-over as the causal phenomenon. The use of precursors for precise analysis of the biosynthetic pathway to anthocyanidin formation helped greatly in identification of the genes involved in this phenomenon (Harrison and Stickland, 1974).

Circumstantial evidence is also available. By analyzing the frequency of spots in *Tradescantia hirsuticaulis,* clone 2, carrying the D^+E^-/D^-E^+ genotype, and by calculating the coefficient of coincidence, Christianson (1975) showed that somatic crossing-over fits as a compatible explanation for the origin of double sectors. In this and another case (Mericle and Mericle, 1974), the twin sectors are extremely likely to be the result of single events.

Vig (1973a) negated the high frequency of occurrence of twin spots in *Glycine* resulting from nondisjunction. The equal, mirror-image components of the twin spots cannot be expected to originate from deletions, gene conversions, or changes in chloroplastic DNA or controlling elements. Besides, chemicals like NaN_3 produce primarily, or only, single spots; and puromycin, daunomycin, fluorodeoxyuridine, actinomycin D, and similar compounds, which do not cause cross-linking of the DNA molecule, do not induce twin spots. Twin spots also are not observed on homozygous genotypes, since reciprocal exchanges become meaningless in producing variations on this background, and unstable genes like *pal rec* are effective only in producing single spots in *Antirrhinum* (Harrison and Carpenter, 1977).

The strongest recombinogen in mitotic cells, mitomycin C, is known to produce reciprocal exchanges involving corresponding positions on homologous chromosomes in human beings (Shaw and Cohen, 1965) and *Vicia faba* (Rao and Natarajan, 1967). This evidence provides clues to the occurrence of similar mechanisms operative in the production of twin spots in mitomycin C-treated materials.

BIOLOGICAL SIGNIFICANCE OF SOMATIC CROSSING-OVER

The study of somatic crossing-over has significance in both pure and applied biology. Perhaps the most important contribution in basic research is the information that can be obtained about the arrangement of chromatin in interphase cells. Because of the very diffuse nature of the material, these cells are not amenable to such studies directly. There is hardly any question about the fact that somatic crossing-over occurs in the period between G_1 and G_2; hence, one can conclude that chromosomes in interphase are paired homologously, at least in certain localized regions (Vig, 1975a). The data from *S. cerevisiae* suggest the involvement of paracentromeric regions (Minet, Grogsenbacher-Brunder, and Thuriaux, 1980), in that the frequency of somatic crossing-over is far higher in this region than in other regions of the chromosome. The studies revealing that in some eukaryotes, chromosomes in mitotic metaphase show somatic association (Antunes-Correia, 1972; Ohno et al., 1961; Kitani, 1963; Chauhan and Abel, 1968; Avivi, Feldman, and Bushuk, 1969) that can be disturbed by applying colchicine may support this notion. The reason that only some species show somatic pairing, or even association, is not clear. The question of known differential occurrence of somatic versus meiotic crossing-over in paracentromeric regions remains unsolved.

Comparative frequencies for a given set of genes can be used to map the proximal or distal positions of the genes. Whether the "relatively high density centromeric DNA among rare mitotic cells" (Minet, Grogsenbacher-Brunder, and Thuriaux, 1980) exists and has any bearing on differential frequencies of mitotic vs. meiotic crossing-over in this region of the chromosome is not known. Also, the assumption that only some mitotic cells are involved in the process of somatic crossing-over contradicts some suggestions that such recombination may be a universal feature of all diploid cells (Zimmermann, Schwaier, and von Laer, 1966). At present we have no direct approach to support either possibility.

Even though G_2 appears to be the most favorable stage for induction of somatic crossing-over, reports of G_1 cells as the target cells are also available (Minet, Grogsenbacher-Brunder, and Thuriaux, 1980). Further studies are needed to show that a lack of a synaptonemal complex (Olson and Zimmermann, 1978) in mitotic cells is really universal. In addition, a clarification of the idea that mitotic cells "transform" into parameiotic cells, with associated localized pairing of chromosome regions to produce somatic crossing-over, and then revert to typical mitotic cells rather than preparing for meiosis (Minet, Grogsenbacher-Brunder, and Thuriaux, 1980), can contribute significantly to the field of chromosome biology.

In applied biology the occurrence and induction of somatic crossing-over can prove useful in mutagenesis, plant breeding, and carcinogenesis. It is known that the efficiency of induction of somatic mosaicism in plants is almost directly related to the efficiency of induction of transmissible mutations (Kiang and Halloran, 1975). Such studies may help to determine the nature of agents efficient in inducing genetic alterations. Suitable concentrations can be determined by study of dose-response curves.

The products of somatic crossing-over can be utilized in obtaining appropriate genetic recombinants for plant breeding by using two approaches. One provides recombination of genes linked in the region proximal to the centromere, since the traditional approach of meiotic recombination is known to have low efficiency for producing recombinants in this region. The second approach is the isolation of appropriate recombinants, in vitro culture of isolates, and regeneration of the whole plant. Both of these approaches are desirable in species where meiotic products have a numerical imbalance of chromosomes or the plant is reproduced only vegetatively—for example, potatoes, sugarcane, or triploids, pentaploids, and so on, as in the families Rosaceae and Rutaceae.

It is generally assumed that the human race has at least five to ten heterozygous gene pairs, in that recessive genes could cause deleterious effects. One such effect is supposed to be the production of cancerous growths in certain homozygous recessive combinations. Somatic crossing-over may bring such genes to expression (Kinsella and Radman, 1978). Also, some agents capable of bringing about this process in higher as well as lower eukaryotes are proven carcinogens.

Examples of carcinogens capable of producing somatic crossing-over can be found in ultraviolet light (Holliday, 1964), bleomycin (Moore, 1978; Hannan and Nasim, 1978; Demopoulos, Stamatis, and Yannopoulos, 1980), several nitrosoamides (Zimmermann, Schwaier, and von Laer, 1966), nitrofuran (Zimmermann and Vig, 1975; Vig and Zimmermann, 1977), nitrous acid (Zimmermann, Schwaier, and von Laer, 1966), and diethyl sulfate (Zimmermann, Schwaier, and von Laer, 1966) (also see other examples from Table 2.5). In addition, concentrations of certain chemicals that produce no mutations or other genetic effects (such as nondisjunction) can bring about somatic crossing-over. Mitomycin C (Holliday, 1964; Vig and Paddock, 1968) and bleomycin (Demopoulos, Stamatis, and Yannopoulos, 1980) are two examples. These data point to the subtle effect of mitotic recombination, induced or spontaneous, and prove a correlation between somatic crossing-over and oncogenesis. This parameter, along with other studies of environmental mutagenesis, deserves better attention, and should not be lost sight of in the study of carcinogenesis.

REFERENCES

Abbadessa, R., and A. B. Burdick. 1963. The effect of X-irradiation on somatic crossing over in *Drosophila melanogaster*. *Genetics* 48:1345-1356.

Ahnström, G., A. T. Natarajan, and J. Veleminsky. 1972. Chemically induced somatic mutations in *Arabidopsis*. *Hereditas* 72:319-322.

Antunes-Correia, J. C. 1972. Somatic association in BHK_{21} homosynkaryons. *Heredity* 28:357-362.

Arenaz, P., and B. K. Vig. 1978. Somatic crossing over in *Glycine max*. (L.) Merrill: Activation of dimethylnitrosoamine by plant seed and comparison with methyl nitrosourea in induction of somatic mosaicism. *Mutation Res.* 52:367-380.

Ashley, T. 1978. Effect of colchicine on somatic crossing over induced by mitomycin C in soybean (*Glycine max*). *Genetica* 49:87-96.

Avivi, L., and M. Feldman. 1973. The mechanism of somatic association in common wheat, *Triticum aestivum* L. IV. Further evidence for modification of spindle tubulin through the somatic association gene as measured by vinblastine binding. *Genetics* 73:379-385.

Avivi, L., M. Feldman, and W. Bushuk. 1969. The mechanism of somatic association in common wheat, *Triticum aestivum* L. I. Suppression of somatic association by colchicine. *Genetics* 62:745-752.

Barrow, J. R., H. Chaudhari, and M. P. Dunford. 1973. Twin spots on leaves of homozygous cotton plants. *J. Hered.* 64:222-226.

Barrow, J. R., and M. P. Dunford. 1974. Somatic crossing over as a cause of chromosome multivalents in cotton. *J. Hered.* 65:3-7.

Becker, H. J. 1956. On X-ray induced somatic crossing over. *Dros. Information Serv.* 30:101-102.

Brosseau, G. E. 1957. The environmental modifications of somatic crossing over in *Drosophila melanogaster* with special reference to developmental phase. *J. Exp. Zool.* 136:367-393.

Carlson, P. 1974. Mitotic crossing over in a higher plant. *Genetical Res.* 24: 109-112.

Chauhan, K. S. P., and W. O. Abel. 1968. Evidence for the association of homologous chromosomes during premeiotic states in *Impatiens* and *Salvia. Chromosoma* (Berlin) 25:297-302.

Christianson, M. L. 1975. Mitotic crossing-over as an important mechanism of floral sectoring in *Tradescantia. Mutation Res.* 28:389-395.

Conger, B. V., and J. V. Carabia. 1977. Mutagenic effectiveness and efficiency of sodium azide versus ethyl methanesulfonate in maize: Induction of somatic mutations at the yg_2 locus by treatment of seeds differing in metabolic state and cell population. *Mutation Res.* 46:285-296.

Cornu, A. 1970. Sur l'obtention de mutations somatiques après traitements de grains de *Petunia. Ann. Amelior. Plantes* 20:189-214.

Cornu, A., and P. Dommergues. 1974. Analyse génétique des modifications induites aux loci A, R, du *Petunia.* In *Polyploidy and Induced Mutations in Plant Breeding*, pp. 63-73. PL503/11. Vienna: IAEA.

Cuany, R. L., A. H. Sparrow, and A. H. Jahn. 1959. Spontaneous and radiation induced mutation rates in *Antirrhinum, Petunia, Tradescantia* and *Lilium. Proc. 10th Int'l. Cong. Genet. Montreal*, vol. 2, pp. 62-63. Toronto: University of Toronto Press.

Dahlgren, K., and V. Ossian. 1927. Ein sektorialchimare vom Apfel. *Hereditas* 9:335-342.

Demopoulos, N. A., N. D. Stamatis, and G. Yannopoulos. 1980. Induction of somatic and male crossing-over by bleomycin in *Drosophila melanogaster. Mutation Res.* 78:347-351.

Deshayes, A., and H. Dulieu. 1974. Études des variations somatiques de deux mutants chlorophylliens de *Nicotiana tabacum* L., leur nature génétique et les facteurs qui les favorisent. In *Polyploidy and Induced Mutations in Plant Breeding*, pp. 85-99. PL 503/14A. Vienna: IAEA.

Dulieu, H. 1972. The combination of cell and tissue culture with mutagenesis for the induction and isolation of morphological or developmental mutants. *Phytomorph.* 22:283-296.

——. 1974. Somatic variations on a yellow mutant in *Nicotiana tabacum.* I. Non-reciprocal genetic events occurring in leaf cells. *Mutation Res.* 25:289-304.

——. 1975. Somatic variations on a yellow mutant in *Nicotiana tabacum*. II. Reciprocal genetic events occurring in leaf cells. *Mutation Res.* 28: 69-77.

Dulieu, H., R. de Boelpaepe, and A. Deshayes. 1971. Sur l' existence spontanée de recombinaisons somatiques chez un mutant de *Nicotiana xanthi n.c.* et leur induction par le rayonnement γ; premières études génétiques. *C. R. Acad. Sci. Paris* ser D. 272:3287-3290.

Evans, D. E. 1977. Modification of the frequency of mitotic crossing-over in *Nicotiana tabacum*, *Glycine max*, and *Lycopersicon esculentum* using x-rays. Ph.D. diss., Ohio State University.

——. 1980. Characterization of x-ray induced increase of mitotic crossing-over in *Glycine max*. *Theor. Appl. Genet.* 56:245-251.

Evans, D. E., and E. F. Paddock. 1979. Mitotic crossing-over in higher plants. In *Plant Cells and Tissue Culture*, ed. W. Sharp, P. O. Larsen, E. F. Paddock, and V. Raghavan, pp. 315-351. Columbus: Ohio State University Press.

Fujii, T. 1980. Somatic mutations induced by furylfuramide (AF-2) in maize and soybean. *Japan. J. Genet.* 55:241-245.

German, J. 1964. Cytological evidence for crossing-over in vitro in human lymphoid cells. *Science* 166:298-301.

German, J., and J. La Rock. 1969. Chromosomal effects of mitomycin, a potential recombinogen in mammalian cell genetics. *Texas Rep. Biol. Med.* 27:409-418.

Gibson, D. A. 1970. Somatic homologue association. *Nature* 227:164-165.

Hannan, M. A., and A. Nasim. 1978. Genetic activity of bleomycin: Differential recombinogen in mammalian cell genetics. *Texas Rep. Biol. Med. Res.* 53:309-316.

Harrison, B. J., and R. Carpenter. 1973. A comparison of the instabilities at the *Nivea* and *Pallidea* loci in *Antirrhinum majus*. *Heredity* 31:309-323.

——. 1977. Somatic crossing over in *Antirrhinum majus*. *Heredity* 38:169-189.

Harrison, B. J., and A. G. Stickland. 1974. Precursors and genetic control of pigmentation. 2. *Heredity* 33:112-115.

Hendrychova-Tomkova, J. 1964. Local somatic colour changes in *Salvia splendens*. *J. Genet.* 59:7-13.

Hirono, Y., and G. P. Rédei. 1965. Induced premeiotic exchange of linked markers in the angiosperm *Arabidopsis*. *Genetics* 51:519-526.

Holliday, R. 1964. The induction of mitotic recombination by mitomycin C in *Ustilago* and *Saccharomyces*. *Genetics* 50:323-335.

Huttner, K. M., and F. H. Ruddle. 1976. Study of mitomycin C-induced chromosomal exchange. *Chromosoma* 56:1-13.

Iyer, V. N., and W. Szybalski. 1963. A molecular mechanism of mitomycin action: Linking of complementary DNA strands. *Proc. Nat. Acad. Sci. U.S.A.* 50:355-362.

Jones, D. F. 1937. Somatic segregation and its relation to atypical growth. *Genetics* 22:489-522.

——. 1938. Translocation in relation to mosaic formation in maize. *Proc. Nat. Acad. Sci. U.S.A.* 24:208-211.

Kiang, L. C., and G. M. Halloran. 1975. Chemical mutagenesis in soybean (*Glycine max* (L.) Merrill) using ethyl methanesulfonate and hydroxylamine hydrochloride. *Mutation Res.* 33:373-382.

Kinsella, A. R., and M. Radman. 1978. Tumor promoter induces sister chromatid exchanges: Relevance to mechanisms of carcinogenesis. *Proc. Nat. Acad. Sci. U.S.A.* 75:6149-6153.

Kitani, Y. 1963. Orientation, arrangement and association of somatic chromosomes. *Japan. J. Genet.* 38:247-256.

Klekowski, E., and D. E. Levin. 1979. Mutagens in a river heavily polluted with paper recycling wastes. Results of field and laboratory mutagen assays. *Environ. Mutagen.* 1:209-219.

Kuhn, E. M. 1976. Localization by Q-banding of mitotic chiasmata in case of Bloom's syndrome. *Chromosoma* (Berlin) 57:1-11.

Mericle, L. W., and R. P. Mericle. 1967. Genetic nature of somatic mutations for flower color in *Tradescantia*, clone 2. *Radiat. Bot.* 7:449-464.

——. 1974. Resolving the enigma of multiple mutant sectors in stamen hairs of *Tradescantia. Genetics* 73:575-582.

Merriam, J. R., R. Nothiger, and A. Garcia-Bellido. 1972. Are dicentric anaphase bridges formed by somatic recombination in X-chromosome inversion heterozygotes of *Drosophila melanogaster*? *Molec. Gen. Genet.* 115:294-301.

Minet, M., A. M. Grogsenbacher-Brunder, and P. Thuriaux. 1980. The origin of a centromere effect on mitotic recombination. *Current Genet.* 2:53-60.

Moore, C. N., and C. P. Haskins. 1935. X-ray induced modifications of flower color in *Petunia. J. Hered.* 26:349-355.

Moore, C. W. 1978. Bleomycin induced mutation and recombination in *Saccharomyces cerevisiae. Mutation Res.* 58:41-49.

Nauman, C. H., A. H. Sparrow, and L. A. Schairer. 1976. Comparative effects of ionizing radiation and two gaseous chemical mutagens on somatic mutation induced in one mutable and two non-mutable clones of *Tradescantia. Mutation Res.* 38:53-70.

Nilan, R. A., and B. K. Vig. 1976. Plant test systems for detection of chemical mutagens. In *Chemical Mutagens, Principles and Their Methods of Detection*, ed. A. Hollaender, vol. 4, pp. 143-170. New York: Plenum.

Nothiger, R., and A. Dubendorfer. 1971. Somatic crossing over in the house fly. *Molec. Gen. Genet.* 112:9-13.

Ohno, S., J. M. Trujillo, W. D. Kapland, and R. Kinosita. 1961. Nucleolus-organizers in the causation of chromosomal anomalies in man. *Lancet* 2:123-26.

Olson, L., and F. K. Zimmermann. 1978. Meiotic recombination and synaptonemal complexes. *Molec. Gen. Genet.* 166:151-166.

Rao, R. N., and A. T. Natarajan. 1967. Somatic association in relation to chemically induced aberrations in *Vicia faba. Genetics* 57:821-835.

Ross, J. G., and G. Holm. 1959. Somatic segregation in tomato. *Hereditas* 46: 224-230.

Rubini, P. G., M. Vecchi, and M. G. Franco. 1980. Mitotic recombination in *Musca domestica* L. and its influence on mosaicism, gynandromorphism and recombination in males. *Genet. Res.* (Camb.) 35:121-130.

Sander, C., R. A. Nilan, A. Kleinhofs, and B. K. Vig. 1978. Mutagenic and chromosome-breaking effects of azide in barley and human leukocytes. *Mutation Res.* 50:67-75.

Serebrovsky, A. S. 1925. Somatic segregation in domestic fowl. *J. Genet.* 16: 33-42.

Shaw, M. W., and M. M. Cohen. 1965. Chromosome exchange in human leukocytes induced by mitomycin C. *Genetics* 51:181-190.

Sparrow, A. H., L. A. Schairer, and R. Villalobos-Pietrini. 1973. Comparison of somatic mutation rates induced in *Tradescantia* by chemical and physical mutagens. *Mutation Res.* 26:265-276.

Stein, O. L., and D. Steffensen. 1959. Radiation induced genetic markers in the study of leaf growth in *Zea. Amer. J. Bot.* 46:485-489.

Stern, C. 1936. Somatic crossing over and segregation in *Drosophila melanogaster. Genetics* 21:625-730.

Szybalski, W. 1965. Chemical reactivity of chromosomal DNA as related to mutagenicity: Studies with human cell lines. *Cold Spring Harbor Symp. Quant. Biol.* (1964) 29:151-159.

Veleminsky, J., and T. Gichner. 1968. The mutagenic activity of nitrosoamines in *Arabidopsis thaliana. Mutation Res.* 5:429-431.

Vig, B. K. 1969. Relationship between mitotic events and leaf spotting in *Glycine max. Canad. J. Genet. Cytol.* 11:147-152.

——. 1971. Increase induced by colchicine in the incidence of somatic crossing over in *Glycine max. Theoret. Appl. Genet.* 41:145-149.

——. 1972. Suppression of somatic crossing over in *Glycine max* (L.) Merrill by deoxyribose cytidine. *Molec. Gen. Genet.* 116:158-165.

——. 1973a. Somatic crossing over in *Glycine max* (L.) Merrill: Effect of some inhibitors of DNA synthesis on the induction of somatic crossing over and point mutations. *Genetics* 73:583-596.

——. 1973b. Somatic crossing over in *Glycine max* (L.) Merrill: Mutagenicity of sodium azide and lack of synergistic effect with caffeine and mitomycin C. *Genetics* 75:265-277.

——. 1973c. Mitomycin C induced mosaicism in *Glycine max* (L.) Merrill in relation to the post germination age of the seed. *Theoret. Appl. Genet.* 43:27-30.

——. 1974. Somatic crossing over in *Glycine max* (L.) Merrill: Differential response to ^{3}H emitted β-particles and ^{60}Co emitted γ-rays. *Radiation Bot.* 14:127-137.

——. 1975a. Chromatin-nuclear membrane attachment in relation to DNA replication and chromosome aberrations: A new hypothesis. *J. Theoret. Biol.* 54:191-199.

——. 1975b. Soybean (*Glycine max*): A new test system for study of genetic parameters as affected by environmental mutagens. *Mutation Res.* 31:49-56.

——. 1977. Genetic toxicology of mitomycin C, actinomycin, daunomycin and adriamycin. *Mutation Res.* 34:189-238.

——. 1978. Somatic mosaicism in plants with special reference to somatic crossing over. *Environ. Health Persp.* 27:27-36.

Vig, B. K., R. A. Nilan, and P. Arenaz. 1976. Somatic crossing over in *Glycine max* (L.) Merrill: Induction of somatic crossing over and specific locus mutations by methylmethanesulfonate. *Env. Exp. Botany* 76:223-239.

Vig, B. K., and E. F. Paddock. 1968. Alteration by mitomycin C of spot frequencies in soybean leaves. *J. Hered.* 59:225-229.

——. 1970. Studies on the expression of somatic crossing over in *Glycine max* (L.) Merrill. *Theor. Appl. Genet.* 40:316-321.

Vig, B. K., and F. K. Zimmermann. 1977. Somatic crossing over in *Glycine max* (L.) Merrill: An induction of the process by carofur, diepoxybutane and trenimon. *Environ. Exp. Botany* 17:113-120.

Zaman, M. A., and K. S. Rai. 1973. Partial floral chimera due to somatic crossing over in *Collinsia heterophylla* Buist, Scrophulariaceae. *Bangladesh J. Bot.* 2:7-9.

Zimmermann, F. K., R. Schwaier, and U. von Laer. 1966. Mitotic recombination induced in *Saccharomyces cerevisiae* with nitrous acid, diethylsulfate and carcinogenic alkylating nitrosamides. *Z. Vererbungsl.* 98:230-246.

Zimmermann, F. K., and B. K. Vig. 1975. Mutagen specificity in the induction of mitotic crossing-over in *Saccharomyces cerevisiae. Molec. Gen. Genet.* 139:255-268.

3

Pollen Mutants and Mutagenesis

Robert A. Nilan
Jeffrey L. Rosichan

INTRODUCTION

The male gametophyte, or pollen, has long been an organ of considerable interest and utility in mutagenesis research, primarily because of its haploid nature. Most of this research, which has produced considerable literature (see reviews by Brewbaker and Emery, 1962; Constantin, 1982; Plewa, 1982), utilizes the pollen as the unit of treatment because of the relatively easily scored end points following radiation or chemical mutagen treatment. These end points include pollen germination and pollen tube growth; chromosome aberrations at pollen and pollen tube mitosis; genetic effects from loss of dominant alleles, especially of genes controlling endosperm development in maize; mutations as revealed in first segregating and subsequent generations; mutations in haploid plants raised from treated pollen (Constantin, 1981; Davies and Hopwood, 1980; Devreux and de Nettancourt, 1974); and seed viability. The objectives of the research are to determine the physiological, genetic, and cytogenetic effects of different types of radiation and chemical mutagens; to elucidate the mechanisms of mutagen action and the nature of genetic and cytogenetic changes; and to induce variants for plant improvement and genetic studies. Lately, pollen germination and tube growth have again become important end points to detect biological activity of environmental pollutants (Ascher, 1981; Bilderback, 1981; Feder, 1981; Pfahler, 1981; Rosen, 1981).

The induction and analyses of pollen mutants constitute a more recent development in pollen mutagenesis (Constantin, 1982; Nilan and Vig, 1976; Nilan, 1978; Nilan et al., 1981; Plewa, 1982). The increasing attention to pollen mutants is in part due to the following:

1. The pollen grain is a product of meiosis, and hence is a functional haploid whose genotype controls the phenotypic expression of its numerous traits and can reveal mutations induced in the sporophytic germ line cells as well as postmeiotically.
2. Because of pollen's size, numbers, and ease of handling, pollen mutants can reveal discrete, low-frequency genetic events in large populations of individuals, and are the only means among higher eukaryotes for providing the degree of genetic resolution found in genetic systems of prokaryotes and lower eukaryotes.

3. Knowledge of the biology and biochemistry of pollen is increasing rapidly.
4. Pollen trait mutants exhibit a dose-dependent response to physical and chemical mutagens.

Pollen mutants have use in at least two areas of mutagenesis and genetics. They are considered prime candidates for developing in situ systems for monitoring and screening environmental mutagens (Constantin, 1982; Nilan, 1978; Nilan, Kleinhofs, and Warner, 1981; Nilan et al. 1981; Plewa, 1978; 1981). In addition, appropriate genetic analyses of pollen mutants can lead to intragenic mapping of alleles (Nelson, 1968; 1976). This genetic fine-structure analysis, combined with chemical analyses of the protein product of the locus in terms of amino acid sequence and peptide mapping, provides an understanding of the nature of mutations (or the type of lesion in DNA) producing the mutant alleles. Moreover, such mutants can help probe the structure, organization, and regulation of the eukaryotic gene (Freeling and Woodman, 1979).

To better exploit pollen mutants in mutagenesis and genetics research, a variety of genetic traits in pollen structure, physiology, and chemistry must be identified and studied. At present very few useful genetic traits have been detected, but this number is growing with increasing knowledge of the biology, especially the biochemistry, of the pollen grain.

The aims of this paper are to summarize information about the biology and biochemistry of the pollen grain and the currently used and potentially useful genetic traits and their mutants, and to appraise the value of pollen mutants for mutagenesis and genetic studies in plants. The latter will include the induction and selection of pollen mutants and the problems of establishing accurate mutant frequencies and rates. In addition, new approaches for measuring nondisjunction (Weber, 1981), chromosome damage (Ma, 1981), and mutations (Mulcahy, 1981) through tetrad analysis will be discussed.

POLLEN BIOLOGY

No discussion of pollen mutants can be meaningful without some knowledge of recent advances in pollen development, structure, and chemistry. The structure and cytology of the pollen grain have been well described (Brewbaker and Emery, 1962; Heslop-Harrison, 1971; Knox, 1981; Maheshwari, 1949). Mature pollen grains are either binucleate (two-celled) or trinucleate (three-celled), with the majority of angiosperm pollen being binucleate. The binucleate consists of a vegetative nucleus to control pollen tube growth and metabolism, and a generative nucleus that divides within the pollen tube to produce two sperm nuclei. The trinucleate tube consists of the vegetative nucleus and two sperm nuclei. The development of the two nuclear states and their consequences in relation to mutagen treatment and to the culture and storage of the pollen have been discussed by Brewbaker and Emery (1962).

For identifying structural features that might have useful variants or mutants, a detailed knowledge of the cell wall and the origin of each layer is essential. Certainly only mutants of traits of gametophytic origin are of value in mutagenesis and genetic studies, since sporophytic traits are of diploid maternal origin.

The structure and development of the pollen grain wall has been described in detail (Heslop-Harrison, 1971; Knox, 1981). It is made up of two layers, the intine and the exine, each having a distinct origin (Figure 3.1). A more detailed view is shown in Figure 3.2 and is described by Heslop-Harrison (1971). The exine is composed

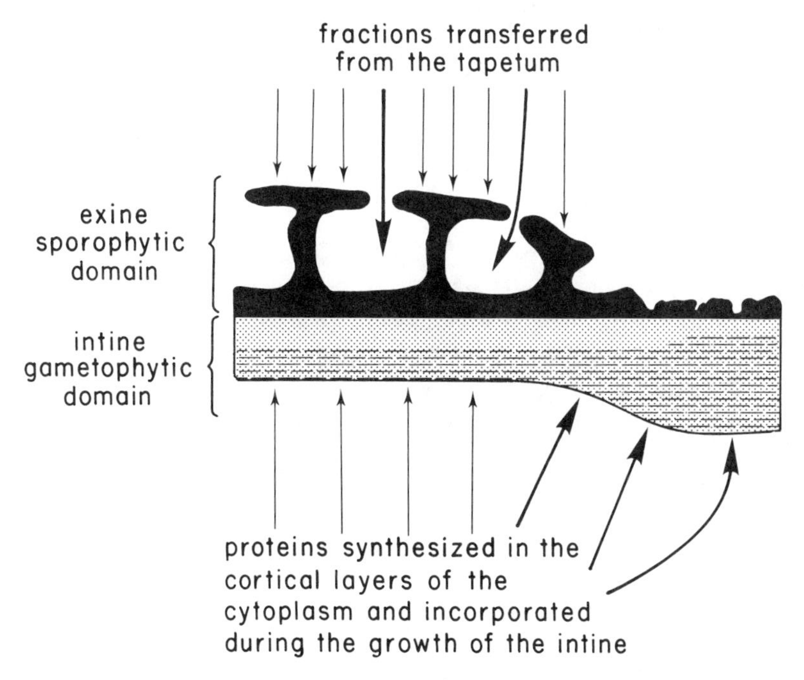

FIGURE 3.1 Diagram of the Pollen Grain Wall Showing the Origin and Storage Sites of the Wall-held Proteins. After Heslop-Harrison (1975a).

CAMROSE LUTHERAN COLLEGE
LIBRARY

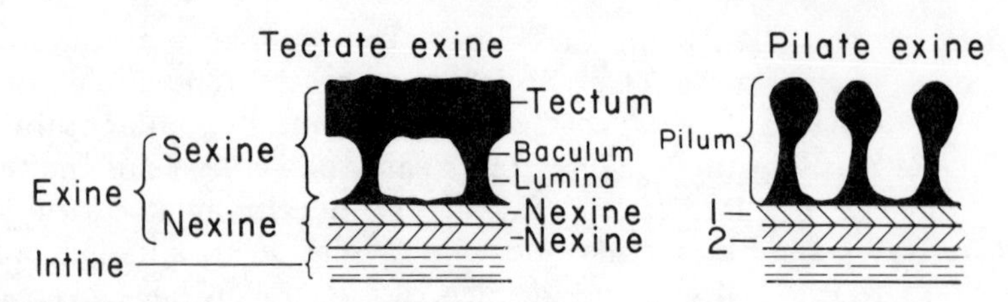

FIGURE 3.2 Diagram of the Pollen Wall Layers. The sexine is made up of a foot layer containing rodlike bacula. The roof of the sexine (tectum) may be perforated by micropores, and is often ornamented with spines and protuberances. In pollen having a pilate exine, the tectum is absent and the tops of the bacula are often swollen. After Heslop-Harrison (1971).

of sporopollenin, considered to consist principally of carotenoids and carotenoid esters (Brooks and Shaw, 1971). The arcades of the exine serve as storage sites for a range of proteins, glycoproteins, carbohydrates, lipids, and pigments (see review by Heslop-Harrison, 1975b). The exine-held fraction is of sporophytic origin and is supposedly produced by tapetal cells and then transferred to the exine late in pollen development (Heslop-Harrison, 1975b). An oily, viscous "pollenkitt" or "tryphine" layer coats the surface of many types of pollen grains.

The intine is gametophytic in origin, and its development, structure, and function are quite well known (Heslop-Harrison, 1971; 1975b; Knox, 1981). It is formed from cellulose, hemicelluloses, and pectic polymers synthesized by the haploid pollen protoplast. Many proteins have been localized to tubules or lamellar structures within the polysaccharide matrix by methods that detect enzyme activity or antigenic activity using fluorescently tagged antibodies. The intine-held proteins are released from around the entire exine or through the germinal pore.

In recent years considerable progress has been made in understanding the biochemistry and molecular biology of the pollen grain (Mascarenhas, 1975; 1981). Particular emphasis has been on identification and analysis of proteins (Brewbaker, 1971; Clarke and Knox, 1978; Heslop-Harrison, 1971; Knox, 1981; Porter, 1981). This work has centered on determining allergen effects, incompatibility relationships ("recognition substances"), various aspects of beekeeping, and the identification of varieties and subvarieties by isozyme patterns (Mäkinen and Macdonald, 1968). From these biochemical analyses is emerging some understanding of the genetics

TABLE 3.1
Enzymes Reported to Be Active in Pollen Grains

Enzyme Group	*Enzyme Trivial Name*
Dehydrogenases	alcohol dehydrogenase
	glucose 6-phosphate dehydrogenase
	glutamate dehydrogenase
	isocitrate dehydrogenase
	lactate dehydrogenase
	lipoamine dehydrogenase
	malate dehydrogenase
	6-phosphogluconate dehydrogenase
	succinate dehydrogenase
	triosephosphate dehydrogenase
	UDPG dehydrogenase
Oxidases	amino acid oxidase
	catalase
	cytochrome oxidase
	peroxidase
Transferases	ADPG glucosyltransferase
	ADPG pyrophosphorylase
	alanine aminotransferase
	aspartate aminotransferase
	aspartate carbonyltransferase
	hexokinase
	nucleoside diphosphate kinase
	phosphoglucomutase
	phosphorylase
	ribonuclease
	sucrose synthetase
	trehalose 6-phosphate synthetase
	UDPG pyrophosphorylase
	UDPG glucosyltransferase
	UDPG:quercetin glucosyltransferase
Hydrolases	acid phosphatase
	alkaline phosphatase
	aminoacylase
	amylase
	cellulase (β-1,4-glucanase)
	cutinase
	esterases
	β-fructofuranosidase (invertase)
	α-glucosidase
	leucine aminopeptidase
	pectinase
	protease (trypsin and chymotrypsin)

(*continued*)

TABLE 3.1 (Continued)

Enzyme Group	*Enzyme Trivial Name*
	trehalase trehalose 6-phosphate phosphatase
Lyases	citrate synthetase ketose 1-phosphate aldolase
Isomerases	glucosephosphate isomerase
Ligases	carboxylases

Source: After Nilan et al. (1981).

of numerous protein traits, including those that could provide useful mutants for antigen, genetic, and mutagenesis studies.

Many of the pollen proteins were described by Brewbaker (1971) and are listed in Table 3.1. Most have not been characterized genetically or localized within the pollen grain. The cell wall proteins, consisting of antigens as well as enzymes, are readily diffusible, with the sporophytic fraction released first (Heslop-Harrison, 1971). It appears that most of the proteins are synthesized and incorporated soon after the release of microspores from the tetrads (Heslop-Harrison, 1971; Mascarenhas, 1975).

INDUCTION OF USEFUL POLLEN MUTANTS

Pollen mutants are induced by treating the zygote, embryo, seedling, developing plant, and plant inflorescence with physical or chemical mutagens up to the end of meiosis. Radiation treatments may be acute or chronic from zygote to inflorescence. Chemical mutagens are delivered through the seed and roots of seedlings or by injection into plants at later stages, including inflorescence. Ingenious and successful methods for chemical treatments of maize tassels have been developed (Amano and Smith, 1965; Neuffer and Ficsor, 1963).

Gamma rays, X rays, neutrons, and ultraviolet light have been the favorite physical mutagens, while several chemicals, particularly ethyl methanesulfonate and other sulfonates and sodium azide, have been utilized. Procedures for conducting controlled experiments with both physical and chemical mutagens in plants are described in the IAEA *Manual on Mutation Breeding* (1977).

Pollen mutants, at least five for waxy (Nelson, 1968), have been induced by controlling elements in maize. Here a genetic element near or in the structural gene inactivates or lowers its expression. Reversions or mutations to partial or full activity occur under the influence of the controlling element.

Mutation induction kinetics are known for pollen mutants. Since most of these studies have been conducted with the waxy pollen locus of maize and barley and the S locus of *Oenothera*, a detailed account will be deferred until later in this paper. In summary, pollen mutant production responds to doses of acute or chronic radiation and to concentrations of chemical mutagens.

A DNA repair system in germinating *Petunia* pollen has been detected as unscheduled DNA synthesis following treatments with ultraviolet and certain chemicals such as 4-nitroquinoline-1-oxide, diethyl sulfate, and N-methyl-N-nitro-N-nitrosoguanidine, which are capable of direct covalent interaction with DNA (Jackson and Linskens, 1978; 1979; 1980). Mutagens requiring metabolic activation before DNA interaction—such as sodium azide, diethylnitrosoamine, and ethidium bromide—did not induce repair synthesis. Hydroxyurea considerably stimulated repair synthesis, and no replicative synthesis occurred in the pollen. Interestingly, mature animal sperm appears not to have a DNA repair capability. Jackson and Linskens (1980) speculate that a repair system is required in pollen because of the relatively long life of pollen and its frequent exposure to ultraviolet irradiation.

SELECTION OF USEFUL POLLEN MUTANTS

Useful pollen mutants are those of genes controlling traits of gametophytic origin. Most important, the traits must be amenable to quick and easy detection—for example, through a stain or an immunological response or clear morphological markers that clearly distinguish the normal from the mutant allele of a given gene. Furthermore, the useful mutants will be those that can be detected at frequencies as low as 10^{-7} and are amenable to scoring in automated counting and recording devices such as flow cytometers and automatic image analyzers (Amano, 1981a; 1981b; Pinkel, 1981; Tyrer, 1981).

The selection of pollen mutants will be aided if the mutant can be detected in another plant part. For instance, changes in the locus controlling starch deposition in maize and barley can be detected in the endosperm, and a plant raised from such a kernel will carry waxy pollen. Similarly, alcohol dehydrogenase (Adh) can be selected electrophoretically, as altered allozymes in the scutellum, and the resulting plants will carry Adh^- pollen (Freeling, 1976). Protocols for selecting these and other mutants will be described below.

MUTANT FREQUENCIES AND MUTATION RATES

Once a pollen mutant has been selected, problems in its utilization in mutagenesis or genetic studies arise. These relate to establishing accurate measurements of its spontaneous reversion frequency, stability, and heteroallelic recombination frequencies, and of the effects of various doses of mutagens, especially at low levels (as experienced from environmental pollutants), on forward or reverse mutant frequencies.

Higher than expected reversion frequencies occur for waxy in maize and barley (Hodgdon et al., 1981; Plewa and Wagner, 1981; Rosichan et al., 1981) and for *Adh*1 in maize (Freeling, 1977). Freeling (1977) found that at the *Adh*1 locus the spontaneous reversion exceeds forward mutant frequencies. He recognized that the comparison of forward mutant and revertant frequencies for specific genes is fraught with "methological pitfalls, hidden assumptions and statistics of dubious applications," and indicated that high reversion frequencies might be expected from insertion-type mutants.

Great variation is exhibited for mutant frequencies of a pollen trait such as waxy. In barley, mutant frequencies vary sometimes tenfold from spike to spike within a single plant. Frequencies between plants are much more uniform, but often show twofold to threefold variation (Hodgdon et al., 1981). This variation is due in large measure to clustering of mutants, and is a complicating factor in establishing accurate mutant frequencies and mutation rates. Clustering of mutants is affected by the developmental stage of the plant at the time of the mutational event, the coincidence of mutational events, the persistence of the mutagen, and the genetically effective cell number (GECN). These and other factors will be discussed briefly, using the barley spike as an example.

The developmental phase at which the mutation occurs greatly influences the cluster size (Lindgren, 1975). A mutation early in the development produces a large mutant sector, while one late in spike development produces a small sector that can be reduced to one or two mutant pollen grains resulting from a mutation at meiosis. The mutant frequency data also are confused by multiple mutation events in one spike (coincidence of mutation events). Since the mutation rate is based on the number of mutation events, the statistical procedure must be able to estimate the amount of coincidence.

Chronic irradiation and persistence or longevity of a chemical mutagen also influence mutant clustering. Recent data have revealed the existence of metabolic activation of chemicals to mutagens in plants (see Vol. 1, Chapters 1 and 2), and these mutagenic metabolites may be long-lived (Owais et al., 1978; Plewa, 1978; Plewa and Gentile, 1976; Scott et al., 1978). A persistent mutagen like azide is more likely to induce multiple mutations at various stages of the spike development than is a nonpersistent mutagen like X rays. Furthermore, it is well documented that mutagen sensitivity varies at different stages of plant development, with the highest at the time of gametogenesis. In many plants the stages of gametogenesis among cells of a single anther are often not synchronous.

The GECN, which is the average primordial cell number for a spike, also affects mutant sector size. The higher the GECN, the smaller the mutant sector will be. The GECN is determined by an analysis of mutant segregation ratios in M_2 and M_3 populations (Li and Rédei, 1969). Table 3.2, which is derived from M_2 segregation ratios of chlorophyll-deficient seedling mutants among 7 different barley lines, shows the variation that can be expected in GECNs. It appears that six-row varieties have a higher GECN than two-row varieties. For proper statistical interpretation, the GECN should be determined, and it should be noted that the mutagen used can affect the GECN.

Methods of statistical analysis with adaptation for pollen mutant frequency data have been published (Hodgdon et al., 1981; Katz,

TABLE 3.2
Genetically Effective Cell Number (GECN) for Various Barley Lines

Line or Variety	*Type*	*M_2 Seedlings Scored*	*Average Seg. Ratio*	*Approximate GECN*
WSU-9044	2-row	6,348	5:1	1–2
WSU-9037	2-row	6,638	6:1	2
Manker	6-row	26,736	10:1	3
Himalaya	6-row	14,285	10:1	3
Advance	6-row	43,734	10:1	3
Steptoe	6-row	39,560	13:1	4
Morex	6-row	28,650	14:1	4+

Note: Data obtained from chlorophyll-deficient mutants among M_2 seedlings.

1978; 1979; Marcus and Czajkowski, 1979). Marcus (unpublished) has performed a large number of computer simulations of the barley pollen mutagenesis process. Models in which the mutagen (persistent) is diluted by the cell number produce great variability in the data, with meiosis introducing even greater variability (including some cases of multimodal density function).

One important factor in stabilizing mutant frequency data is consistent sampling of primary spikes from the barley plants. This method permits sampling of primordial cells (target) that have received a uniform mutagen treatment.

After analysis of the variability of pollen mutant frequency data, a reasonable sampling method can be developed. Our method for spike analysis in barley involves using a randomized plot design with three or four replications per treatment. From each replication (plot) 15 plants are chosen at random, and five or more spikes are harvested from each plant. The four spikes with the highest pollen count, or best pollen development, are scored for mutation frequency.

POLLEN GENE AND CHROMOSOME MUTANTS

As indicated previously, an extensive review of the literature has determined that at present very few pollen traits whose mutants appear useful for mutagenesis and genetic studies are available. This section will provide and appraise current knowledge about these few well-known traits. Moreover, it will describe all currently known traits and indicate those whose mutants appear to have some value. Among the traits reviewed are inviability (including male sterility), incompatibility, proteins, and enzymes and starch deposition. In addition, the use of certain chromosome mutants in pollen will be described.

It would seem that the well-known variations—and these are quite numerous—in shape and ornamentation or sculpturing of the outer layers (tectum) of the outer wall (exine) that can be readily observed under a light microscope might be considered useful genetic traits for mutagenesis studies. Sculpturing or ornamentation patterns of the outer wall involve the distribution and dimensions of spines, ridges, perforations, papillae, and patterns of spinulae on ridges. These various shapes and ornamentations and their potential use in taxonomy have been discussed (Erdtman, 1952; Maheshwari, 1949). Of importance here is the fact that these patterns and shapes that might appear to be useful pollen genetic markers are of sporophytic origin, and thus are of no value in the context of this review. They are mentioned here only because it is assumed that many readers would consider such patterns to be readily adapted to genetic and mutagenesis studies.

Pollen Inviability

Pollen inviability, especially genetic male sterility, has been detected in nature or following mutagen treatments in a wide variety of plants. Many loci are involved, and should constitute a sensitive means for detecting and measuring effects of environmental mutagens (Mulcahy, 1981; Pfahler, 1975). For instance, 16 (Hockett and Eslick, 1968; 1971), 14 (Coe and Neuffer, 1977; Neuffer and Coe, 1974), and 12 (Rick, 1974) loci for genetic male sterility are known in barley, maize, and the tomato, respectively.

Cytoplasmic male sterility has also been uncovered in a number of vegetable crops, maize, wheat, and (more recently) barley. This too leads to aborted pollen, thus making the scoring for forward mutations of genetic male sterile loci complicated. However, reverse mutations in a homozygous genetic male sterile line would lead to normal pollen, which could be detected by a vital stain such as fluorescein diacetate (Heslop-Harrison and Heslop-Harrison, 1970) and through seed set. Both of these could be readily scored for frequencies of reverse mutations, which might be useful in mutagenesis and genetic studies.

In most plant species it may be difficult to distinguish unambiguously mutation-induced inviability from nonmutagenic environmentally induced pollen abortion. The use of pollen shed as tetrads has been proposed to overcome this obstacle (Mulcahy, 1981). Normally the callose wall separating the four meiocytes breaks down prior to anthesis in flowering plants. Forty-one families in which mature pollen is shed as dyads, or more commonly as tetrads, have been reported (Erdtman, 1945); and ones that may be well suited for studying mutagen-induced inviability have been discussed (Mulcahy, 1981). Mutation-induced inviability would be expected to show genetic segregation for viable and inviable grains within the tetrads. Conversely, nonmutagenic environmental stresses would lead to large numbers of nonsegregating tetrads containing aborted grains.

Several methods, as well as their advantages and limitations, for testing pollen viability in tetrads have been discussed (Mulcahy, 1981). The simplest involves the use of 0.05 percent aniline blue dissolved in lactophenol to test for the presence of cytoplasm in pollen grains (Darlington and La Cour, 1969). The use of aniline blue may overestimate viable pollen, since not all pollen-containing cytoplasm may be viable, thereby underestimating mutant frequency. The fluorescein diacetate test (Heslop-Harrison and Heslop-Harrison, 1970) may provide a more accurate estimate of viability, since metabolism is required to produce fluorescence. Simultaneous counterstain by 0.2 percent trypan blue (Melamed, Kamentsky, and Boyse, 1969) may further enhance the sensitivity, since illumination by ultraviolet light with a weak background of visible light causes viable grains to fluoresce bright yellow-green, while inviable grains stain black. The use of fluorescein requires immediate scoring of pollen, since the stain is lost as the cells die. Size and/or volume differences between viable and inviable grains also may be used to distinguish them from one another.

Incompatibility

Incompatibility, the inability of a plant capable of functional gamete production to set seed when self-pollinated, represents a much-studied pollen mutant system that has been used in genetic and mutagenesis studies. Its genetics and biochemistry have been reviewed (Clarke and Knox, 1978; Heslop-Harrison, 1975b; Knox, 1981; Mulcahy and Johnson, 1978).

Two types of incompatibility are known, sporophytic and gametophytic. In sporophytic incompatibility the mating behavior of the pollen is determined by the genotype of the pollen parent. In gametophytic incompatibility the mating behavior is determined by the pollen genotype. This discussion will be limited primarily to gametophytic incompatibility.

The studies of Lewis (1947; 1949; 1960) using *Oenothera organensis* have produced considerable genetic knowledge of incompatibility, and provide the best "characterized" example. Incompatibility in *Oenothera* is gametophytic and controlled by a single locus, *S*, having more than 50 alleles. The gene is postulated to consist of three closely linked cistrons, one each controlling *S* specificity, activation in the style, and activation in the pollen (Lewis, 1960). Lewis (1960) suggests that the mutants are due to a single mutation event. Several models have been proposed (Brewbaker and Natarajan, 1960; de Nettancourt et al., 1975; Pandey, 1967) that would require at least two mutation events to create a chromosome fragment carrying an extra *S* allele believed to be responsible for mutant formation.

The mechanism of sporophytic incompatibility appears to involve the exine-held proteins/glycoproteins (Dickinson and Lewis, 1973; Heslop-Harrison, Knox, and Heslop-Harrison, 1974; Heslop-Harrison et al., 1975; Knox, Willing, and Ashford, 1972), and possibly pollen antigens (Lewis, Burrage, and Walls, 1967). *Oenothera organensis* pollen of genotype S_2S_3 placed on agar containing genotype S_6 antiserum formed no precipitin rings, while S_6S_6 and $S_6'S_6'$ pollen formed precipitin rings, and approximately 50 percent of the pollen from S_3S_6 plants was found to form precipitin rings on agar containing S_6 antisera. There is no experimental evidence demonstrating a role of intine-held proteins in gametophytic incompatibility responses (Clarke and Knox, 1978) and, despite the body of genetic and biochemical knowledge, the precise nature of the *S* gene product and its mechanism of action remain subject to speculation.

Like all pollen systems for detecting mutagen effects, incompatibility has the advantage that pollen, being haploid, will allow for ready detection of recessive mutants. The incompatibility system offers the added advantage that mutants can affect fertilization, allowing recovery and further testing (Devreux and de Nettancourt, 1974). Finally, since the screening is done by the plants, this eliminates the need for great skills on the part of technicians, and allows large amounts of pollen to be easily screened per pollination.

The present limitations of incompatibility for monitoring environmental mutagens have been discussed (Mulcahy and Johnson, 1978). *Oenothera* and *Prunus* were shown to be not as sensitive to irradiation as the *Tradescantia* stamen hair system.

Previously there had been no effort to maximize the mutagen sensitivity of incompatibility systems. However, studies by Sree Ramulu, Schibilla, and Dijkhuis (1981) showed that *Oenothera organensis* can be used to detect the effects of low doses (2.5–20 rad) of fast neutrons and X rays, as measured by increased frequency of *S* mutations, if appropriate clones are used.

Mulcahy and Johnson (1978) point out three criteria necessary for developing incompatibility systems to assay environmental mutagens. First, species that exhibit a high frequency of induced mutations to self-compatibility must be identified. Second, within these species, clones that show a high degree of incompatibility should be selected. Third, techniques must be developed for employing the selected clones for in situ monitoring. They suggest that grasses containing two loci for gametophytic incompatibility (Lundquist, 1954; 1975), *Ranunculus acris* with three loci (Østerbye, 1975), and *Beta vulgaris* with four loci (Larsen, 1977) may be better suited for detection of low doses because more loci provide greater sensitivity to environmental mutagens. Finally, they outline a method for the development of an autonomous in situ assay system that would include electrophoretic markers, allowing the effects of a large mutant sector to be distinguished from the effects of a high mutation frequency in a plant.

Pollen Proteins

Enzymes

Most proteins have not been localized within the pollen grains or characterized as to their genetic control. Therefore, much more work is necessary before their potential use in mutagenesis and genetic studies can be evaluated. Nevertheless, several pollen proteins described below have some promise. They are gametophytic and nuclear in origin; activity-stainable or immunologically identifiable in intact pollen grains; and the product of one to three genetic loci. Such characteristics facilitate detection and screening of genetic variants.

Acid Phosphatase. Acid phosphatase is the most widely studied of the pollen proteins (Efron, 1970; 1971; 1973; Knox, 1971; Vithanage and Knox, 1979). It is localized in the intine (Heslop-Harrison and Heslop-Harrison, 1973; Knox and Heslop-Harrison, 1970; Vithanage and Knox, 1979), and controlled by at least three unlinked loci, AP_1, AP_2, and AP_3, in maize (Efron 1970; 1971; 1973). The enzyme activity is stainable by incubating pollen for one to five minutes with α-naphthyl acid phosphate coupled with hexazonium pararosanilin to give a reddish-brown color (Barka and Anderson, 1962). The substrates naphthol AS-B1 or AS-TR phosphates can also be used; however, the reaction is not as rapid (Knox and Heslop-Harrison, 1970). In addition, fast garnet GBC can be used as a coupler to yield a purple-black color, and has been shown to specifically inhibit AP_3 (Efron, 1969).

Three AP_1 isozymes occur in maize pollen (Efron, 1970). The enzyme appears to be a dimer, and is synthesized after meiosis, since pollen from heterozygous plants lacks hybrid enzyme. Segregation ratios of AP_1 isozyme activity levels indicate that the isozymes are controlled by a single gene.

Leucine Aminopeptidase. Leucine aminopeptidase is widely distributed in angiosperm pollen (Mäkinen and Macdonald, 1968). In maize endosperm there are apparently four genes involved (Beckman, Scandalios, and Brewbaker, 1964). However, in pollen it appears that only one to three are expressed for a given species (Brewbaker, 1971; Mäkinen and Macdonald, 1968). Genetic polymorphisms occur in maize pollen for the LP_2 locus (Brewbaker, 1971). Electrophoretically separated proteins are activity-stained, using L-leucyl-β-naphthylamide HCl as a substrate and Black K salt, a diazonium dye, as a coupler, to yield a deep purple color. Such a procedure might be useful for pollen staining, but the enzyme has not yet been localized.

Alpha-Amylase. Alpha-amylase (Amy) is apparently under the control of a single gene in maize (Nelson and Burr, 1973) and barley (Frydenberg and Nielsen, 1966; Frydenberg, Nielsen, and Sandfaer, 1969). In barley there are at least three alleles at the *Amy* locus that give rise to the three isozymes alpha-type 1, alpha-type 2, and alpha-type 3. Jones and Chen (1976) used immunofluorescent antibodies to localize alpha-amylase in barley aleurone cells. It should be possible to use a similar method for detecting alpha-amylase in pollen. Amylase has been shown to be readily diffusible from birch pollen after one minute, indicating that it may be sporophytic in origin (Laine, 1977). However, Knox and Heslop-Harrison (1970) report that the activity was localized in the intine.

Alcohol Dehydrogenase. Alcohol dehydrogenase (Adh) has been studied most extensively in maize, where it is controlled by two unlinked genes, *Adh*1 and *Adh*2 (Freeling, 1976; 1978a; Schwartz, 1971a; 1971b; 1975). The functional enzyme is found electrophoretically as either a homodimer or a heterodimer of the gene products of the *Adh*1 and *Adh*2 loci. Only *Adh*1 is expressed in the pollen grain, and most of the Adh polypeptides are synthesized after anaphase II of meiosis (Schwartz, 1971b). Adh is inducible, and the activity can be greatly increased with anaerobic stress (Freeling and Schwartz, 1973). This increase in activity has been shown to involve de novo synthesis of Adh mRNA (Ferl, Brennan, and Schwartz, 1980). Numerous mutants induced with ethyl methanesulfonate (EMS) have been characterized by electrophoretic mobility, antigen properties, level of enzyme activity, thermal stability, and phytic acid affinity (Schwartz, 1981). These mutants may be due to point mutations or very minute deletions (Schwartz, 1981). Freeling and Cheng (1978) report the induction of deletion-induced regulatory mutants following radiation treatment and allyl alcohol selection (Schwartz and Osterman, 1976) of *Adh*-null or low-level *Adh* mutants. They indicated that the relative biological efficiency (RBE) for neon-20 was about 5 for mutational events of the *Adh* locus.

Because of the easy accessibility of the gene product and the large amount of maize pollen that can be collected and screened with relative ease, this system perhaps offers the most potential for mutagenesis and genetic studies. The use and potential of maize *Adh*1 as a monitor of environmental mutagens have been discussed by Freeling (1978b; 1981) and Schwartz (1981), and will be discussed further in a later section.

Antigens and Allergens

A large number of antigens have been detected in grass pollens, using crossed radio-immunoelectrophoresis (Weeke and Lowenstein, 1973). Both grass and ragweed allergens have been shown to cross-react with immunologically similar fractions in related pollen (Lee and Dickinson, 1979). However, some antigens, like antigen A of *Lolium perenne*, show specificity only for closely related genera (Watson and Knox, 1976). The effects of mutagens on the pollen antigens have not been studied.

There are several pollen antigens that may be useful for mutagenesis and genetic investigations (Table 3.3). Of these, antigen E is perhaps the best-characterized. King, Norman, and Connell (1964) isolated this antigen from *Ambrosia elatior* (ragweed), and found it to be a protein of molecular weight 37,000 daltons, made up of four subunits. Immunofluorescent antibodies (Knox and Heslop-Harrison, 1971) revealed that antigen E of *Ambrosia* is localized in the intine. Antigen E was found to be present in both the intine and the exine of *Cosmos* pollen (Howlett, Knox, and Heslop-Harrison, 1973). Later, Knox, Heslop-Harrison, and Heslop-Harrisson (1975) demonstrated that small quantities of antigen E are held in the cavities of the exine while large amounts are present in the intine. They propose that in pollen possessing a permeable exine, losses from this layer may be accounted for by fractions passing through it. This may explain some of the confusion about the precise localization of antigen E. In addition, the antigens are freely diffusible from fresh pollen; therefore, it may be necessary to use frozen pollen for effective immunofluorescent detection and screening.

TABLE 3.3
Properties of Some Purified Glycoprotein Allergens from Grass and Ragweed Pollen

Pollen	*Fraction*	*Method of Isolation*	*Percent N (Percent Protein)*	*Percent Carbohydrate*	*MW*
Lolium perenne	group I	D, DC, GF	13	5	27,000 (acidic)
	group I	SP, D, IEC, GF, GLC	(95)	5	33,000 (acidic)
	antigen A	SP, D, IEC, GF, GLC	(90)	10	68,000 (basic)
Phleum pratense	antigen B	SP, EP, D, DC	14	13	10,500 16,000 (acidic)
Ambrosia elatior	antigen E	SP, GF, DC	17	0.2	37,800 (acidic)
	antigen K	SP, GF, DC	17	0.6	38,200 (acidic)
	antigen Ra3	SP, GF, DC, IEC	18	7	11,000 (basic)

Note: Sp = salt precip.; DC = DEAE cellulose chromatography; GF = gel filtration; D = dialysis; EP = ethanol precip.; IEC = ion exchange chromatography; GLC = gas liquid chromatography.

Source: After Knox (1981).

Several pollen antigens have been purified and characterized (Table 3.3). The majority are acidic proteins or glycoproteins with molecular weights ranging from 11,000 to 68,000 daltons. The major allergens of ragweed antigen E and grass pollen group I allergen appear to be present in multiple electrophoretic forms (Knox, 1981). Enzyme activities associated with partially purified allergen preparations also have been reported (Knox, 1981).

Starch Deposition

Mutants of two starch deposition loci, waxy (*wx*) and amylose extender (*ae*), can be detected in pollen. The waxy trait has been observed in over 15 genera of angiosperms, and is characterized by failure of amylose deposition in the endosperm and pollen grain starch (Eriksson, 1969). Starch granules in the cereal pollen grain begin to appear just after the second pollen mitosis (Clapham, 1977), and are gametophytically controlled (Eriksson, 1969). Waxy pollen and nonwaxy pollen stain differentially with iodine, blue for pollen with normal starch and amylose content, and reddish-brown for pollen with amylopectin and no amylose. All of the early genetic research related to waxy pollen has been reviewed by Eriksson (1969). Genetic studies in maize, barley, rice, and sorghum reveal that the waxy character is inherited as a monofactorial recessive.

Waxy mutants have been induced by radiation and chemicals. Stadler (1928) and Konzak and Singleton (1956a; 1956b) indicated that waxy mutants induced by irradiation of maize pollen were due to chromosome deficiency. Many mutations in maize were induced by ethyl methanesulfonate (Plewa and Wagner, 1981), and were transmitted to the second generation and were correlated with less cytogenetic damage than those induced by neutrons (Amano, 1968; 1972; Amano and Smith, 1965). Among the chemical mutagens, sodium azide (Nilan et al., 1981), EMS, and isopropyl methanesulfonate (Lindgren and Eriksson, 1971) have been used to induce waxy mutants in barley.

Several experiments have determined the dose effects of radiation on the gene controlling the waxy trait in maize and barley. In these experiments the mutants were scored in pollen grains, and it is clear that the waxy locus in barley and maize is responsive to mutagenic treatment. Following chronic radiation, both linear and curvilinear or exponential responses were obtained for forward and reverse mutation (Bianchi, 1965; Briggs and Smith, 1965; Ehrenberg and Eriksson, 1966; Eriksson, 1962; 1963). Radiation doses of 3 to 5 R significantly increased mutational events at the waxy locus above control. However, genetic effects of doses as low as 1 rad can be detected, according to Ehrenberg and Eriksson (1966). de Nettancourt et al. (1977) determined that RBE of fast neutrons for mutational events at the waxy locus is lower than 3. In several experiments (Ehrenberg and Eriksson, 1966; Eriksson, 1963; 1971; de Nettancourt et al., 1977), low doses had a greater influence than medium doses in inducing mutational events. This was attributed to the possibility that lower doses do not evoke repair systems, as medium and higher doses do. It was recognized, however, that in the chronic radiation experiments, changes in sensitivity during the development of the plant could occur that might confound the ultimate mutant frequency.

Amano (1981b) has reported the presence of intermediate or "leaky" waxy alleles in maize following EMS treatment. These alleles were found to map throughout the waxy locus, and he suggests that they represent mis-sense mutations able to produce normal-size but partially inactivated enzymes.

The waxy locus of maize is on chromosome 9 between bronze (*bz*) and viridis (*v*), and has been studied extensively by Nelson (1962; 1968; 1975). At present 31 different waxy alleles have been mapped at this locus (Plewa and Wagner, 1981). Five of these were induced by controlling elements (Nelson, 1968).

Genetic studies demonstrate that the waxy locus (*glx*) in barley is on the short arm of chromosome 1 (Nilan, 1964). The two closest linked genes are the chlorina gene (vir^3), approximately 18 map units distal to *glx*, and the gene for virescent seedling (*yc*), approximately 20 map units proximal to *glx* with respect to the centromere. Thirty mutants have been induced by sodium azide and gamma rays, and they map into at least nine allelic groups (Nilan, Kleinhofs, and Warner, 1981).

The waxy locus offers several distinct advantages for genetic and mutagenesis studies. Because mutants are identified by iodine staining, the procedure is relatively simple and not as time-consuming as procedures requiring cumbersome electrophoretic detection methods. In addition, the pollen screening procedure is much less costly than those requiring complex reagents or immunochemicals for mutant identification.

The biochemistry of the waxy gene product has been investigated. Waxy mutants in maize completely lack starch granule-bound uridine diphosphoglucose (UDPG) glucosyltransferase activity (Nelson and Rines, 1962), and enzyme activity is linearly proportional to the number of wild-type (*Wx*) alleles present in the endosperm (Tsai, 1974). These results support the suggestion that the waxy locus is the structural gene for the UDPG glucosyltransferase. Nelson and Tsai (1964) found that maize waxy mutants actually contained about one-tenth the level of bound starch synthase activity compared with normal maize. Nelson, Chourey, and Chang (1978) showed that this low activity could be accounted for by the presence of a second bound starch synthase that had higher affinity for the substrate. They postulated that this second starch synthase could be involved in primer formation.

Amylose-extender mutants of maize have received relatively little attention compared with the waxy mutants. Pollen also can be stained for mutants of the amylose-extender (*ae*) locus, using a modified iodine staining procedure (Moore and Creech, 1972). This procedure has been used to map five alleles of the amylose-extender locus (Moore and Creech, 1972). Amylose-extender mutants are characterized by a lowered level of starch-branching isoenzyme IIb activity, normally associated with soluble starch synthase I (Preiss and Levi, 1980).

Chromosome Mutants

Pollen tube mitosis and meiotic cells have been used extensively to detect spontaneous and induced chromosome mutants or aberrations. This topic has been reviewed (Ehrenberg, 1971; Nilan and Vig, 1976; also see Chapter 1 of this volume), and will not be discussed in this section. However, two systems are being developed, utilizing tetrads and pollen, that show potential for use in mutagen screening and in situ monitoring. These are the micronucleus assay in *Tradescantia* (Ma et al., 1978; Ma, 1979; 1981) and nondisjunction in maize (Weber, 1981).

The micronucleus test has long been used in mammals as an assay for chromosome damage, and with recent technical advances has proved to be more sensitive than standard aberration analysis (Schmid, 1975; Tates et al., 1980). In plants, however, this test has received little attention. Ma and co-workers (1978) first reported on a novel system using tetrads in *Tradescantia* to detect micronuclei induced by 1,2-dibromoethane. Since then this system has been shown to be more sensitive than standard chromosome techniques, and the utility of this system has been discussed (Ma, 1979). It appears to lend itself to large-scale testing because of the ease in handling and its technical simplicity.

Test systems to detect nondisjunction that employ higher eukaryotic plants and animals are cumbersome, inefficient, and expensive, or the event indicative of nondisjunction may be ambiguous (Zimmermann et al., 1979). In maize, two systems are being developed that may overcome this ambiguity (Weber, 1981): one utilizing the nucleolar organizing region (NOR) of chromosome 6 and a second utilizing a complementing null mutation assay. Weber (1981) has suggested that a good candidate for the latter system would be the alcohol dehydrogenase-1 (*Adh*1) locus, since most of the techniques have already been developed for analysis of mutant and normal pollen.

USES OF POLLEN MUTANTS

As indicated above, mutants of pollen traits provide a powerful and sensitive system with high genetic resolution for measuring effects of mutagens and for elucidation of the nature of mutations and the organization and structure of the eukaryotic genome. Combining this system with the relatively long growth period (for chronic exposure to test agents) and wide adaptation of certain plants can lead to useful in situ environmental mutagen monitoring systems.

In Situ Mutagen Monitoring

Monitoring the genetic hazards in the human environment is an increasingly important activity in mutagenesis research. The advantages of plants and pollen genetic systems for in situ monitoring of environmental mutagens have been described (Constantin, 1978; 1982; Ehrenberg, 1971; Nilan, 1978; Nilan and Vig, 1976; Nilan et al., 1981; Plewa, 1978; Plewa and Wagner, 1981; de Serres, 1978). For the human such test systems can monitor real or potential hazards in and around sources of air pollution (factories, laboratories, reactors, chemical waste dumps), with mobile air pollution testing laboratories. For the plants such assays also can detect and predict increases in genetic variability that lead to decreases in genetic purity of food crop cultivars.

Procedures for processing pollen grains for mutant analysis in large-scale tests are modified from those of Nelson (1962; 1968), and are described for maize (Plewa, 1978) and barley (Rosichan et al., 1981). These include pollen analysis methods for measuring either forward (*Wx* → *wx*) or reverse (*wx* → *Wx*) mutations. Automated methods for screening mutant pollen have been (Amano 1981a; 1981b) or are being developed (Pinkel, 1981; Tyrer, 1981). These are based on flow cytometry and image analyses.

Suitable statistical analyses for measuring reverse or forward mutant frequencies and mutation rates were described earlier, and can be found in more detail in Hodgdon et al. (1981) and Plewa and Wagner (1981). Ehrenberg (1977) has reviewed the statistics required for mutagen testing assays.

The most thoroughly developed pollen mutagen monitoring system is waxy maize (Gentile, Wagner, and Plewa, 1978; Plewa and Gentile, 1976; Plewa and Wagner, 1981). Plewa and colleagues use inbreds of maize, homozygous *wxwx* (for reversion frequencies), to detect the mutagenic properties of numerous pesticides or combinations of pesticides in the field under modern farming conditions. They have also used this system to evaluate the mutagenicity of Chicago sewer sludge. To make the system more useful, a very early-maturing dwarf type of plant that might eventually be utilized in mobile mutagen monitoring laboratories (Schairer et al., 1978) is being developed (Plewa and Wagner, 1981).

The waxy locus in maize has produced positive results when testing for mutagenic activity in the atmosphere adjacent to a lead smelter in southeastern Missouri (Lower, Rose, and Drobney, 1978). The waxy locus of barley has been used to measure the effects of various concentrations of ethylene oxide in a gas chamber (Sulovská et al., 1969). Very low concentrations (12.5 ppm) were found to increase the mutant frequency. This barley system was also used in Stockholm to measure the mutagenicity of car exhausts and city air (Lindgren and Lindgren, 1972). However, no strong mutagenic effect was observed. It also recorded positive genetic effects of low doses of radiation at various distances from a CS-137 source from the French Atomic Energy Establishment at Cadarache, France (Faure, Delpoux, and Dalebroux, 1981).

A waxy pollen system in barley for mutagen monitoring (both frequencies and nature of mutation events) is being developed (Nilan, Kleinhofs, and Warner, 1981; Nilan et al., 1981). Appropriate waxy pollen mutants are being selected from among numerous mutants that have been induced by gamma rays, ethyl methanesulfonate, and sodium azide.

The alcohol dehydrogenase-1 (*Adh*1) gene of maize has been proposed as a monitor of environmental mutagens by Freeling (1978b; 1981) and by Schwartz (1981). This system brings together high genetic resolution, a mutation selection scheme, and biochemical analysis, and can measure the effect of mutagens on both structural and regulatory genes. Indeed, Freeling believes that birth defects and cancer may be mediated preferentially by lesions in regulatory DNA, and any mutagen monitoring system must include such analyses. Effective selection procedures for selecting Adh-negative pollen grains have been developed (Freeling, 1978b; 1981; Freeling and Cheng, 1978; Schwartz and Osterman, 1976). To date, over 100 EMS-induced mutants of Adh1 have been selected, and appear to be lesions in the structural gene (Schwartz, 1981).

In utilizing a plant system such as waxy for detecting environmental mutagens and evaluating the biological hazards to humans and plants, it is important to be aware of plant-mediated activation (Plewa and Wagner, 1981; Vol. 1, Chapters 1 and 2). This area of plant mutagenesis was mentioned earlier, and was reviewed by Plewa and Gentile (1981).

Intragenic Mapping and Analysis of the Nature of Induced Mutations and Gene Organization and Structure

It was stated earlier that the pollen mutants provide probably the most powerful means at present among eukaryotic genetic systems for analyzing genetic fine structure through intragenic mapping and genetic and biochemical analyses of mutant alleles. Such analyses can provide some information on the nature of mutations and the structure and organization of the eukaryotic gene. The kinds of mutational events that occur in plants have been described in detail (Freeling, 1978b; Freeling and Woodman, 1979; Nilan, 1967; 1978; Nilan and Vig, 1976).

The genetic fine structure of the waxy locus, through intragenic recombination and mapping, was first analyzed by Nelson (1959; 1962; 1968). His findings, followed by those of Briggs, Amano, and Smith (1965), Amano (1968), and Briggs and Smith (1965), have identified 31 different alleles in the waxy locus in maize. Furthermore, EMS-induced mutations within the locus involve sites different from those induced by fast neutrons. Nelson (1975) demonstrated nonreciprocal or gene conversion following analysis of recombination between two waxy heteroalleles. Twenty-one *Adh*1 mutants have been tested for ability to recombine intragenically, and all behave as unique points (Freeling, 1976; 1978b; Freeling and Woodman, 1979).

It appears that most pollen mutants are the result of intragenic chromosome abnormalities. Certainly there is still insufficient evidence to support DNA base changes as the origin of the mutants. An exponential relation between mutation frequency and dose for radiation-induced reversions at the waxy locus in barley suggested that the mutations were caused by both point mutations and chromosomal aberrations, possibly deletions (Eriksson, 1962). From analyses of radiation-induced forward and reverse mutations at the waxy locus in maize, Bianchi (1965) concluded that although the spontaneous mutations at this locus may be intragenic, certainly the induced mutations are probably due to chromosome aberrations. After a very thorough analysis of 70 radiation-induced mutants of the *Adh*1 locus of maize, Freeling and Cheng (1978) concluded that all the mutants were due to chromosome changes. No intragenic lesions were recovered, although the experimental design would have preferentially recovered them if they had occurred.

The fact that reversions have been induced by Eriksson and

others would suggest that at least deletions are not produced to a great extent by mutagens at the waxy locus. However, as Freeling (1981) indicates, these may suggest insertion elements. Insertion elements are known in the maize waxy locus (Nelson, 1968).

Freeling and Woodman (1979) and Freeling and Cheng (1978) probably provide the best analysis of how pollen mutants such as those of *Adh*1 might reveal the organization of the eukaryotic genome, and the relation of structural to regulatory genes in the development of traits. They suggest that structural genes probably comprise but a small percentage of the nuclear genome of any higher organism. They also point out the inadequacies of determining gene structure by intragenic mapping only, without a detailed analysis of the protein product of the gene locus.

CONCLUSIONS

Recently there has been renewed interest in mutants of pollen traits. The high genetic resolution possible with pollen mutant alleles, combined with biochemical analysis of pollen mutant proteins, can provide insights into the nature of mutations and the structure and organization of genes. In the area of mutagenesis, pollen mutants have particular value in the development of environmental mutagen monitoring systems. The increased utilization of pollen mutants in these research areas has been aided by increased knowledge of the chemistry and structure of the pollen grain.

Pollen mutants are easy to induce by physical or chemical mutagens, and with good screening techniques are easy to select. Mutations at loci exhibit linear or exponential response to these mutagens. Useful pollen traits are gametophytic in origin, have mutants that are detected at frequencies as low as 10^{-7} and are amenable to scoring in automated counting and recording devices.

One problem is establishing reliable forward and reverse spontaneous mutant frequencies, because mutational events can occur at any point from embryo to meiosis. Furthermore, translating mutant frequencies into mutation rates is difficult. A computer-simulated and -assisted statistical analysis may establish more-reliable frequencies and rates.

Mutants suitable for use in mutagenesis and genetic studies may be found among genes controlling viability (male fertility), proteins (enzymes and allergens), incompatibility, and starch deposition. A number of proteins have been identified, although most have not been localized within the pollen grain or characterized as to their genetic control. Among the enzymes, alcohol dehydrogenase (Adh1) is an elegant system that combines genetic analysis unparalleled in eukaryotes with a rigorous biochemical analysis of mutant alleles that can make valuable probes of the eukaryotic genome and has potential as a mutagen monitor. Other enzymes, such as acid phosphatase and leucine aminopeptidase, and certain allergens with very specific detection techniques, have some promise in these areas of study. Incompatibility (*S*) mutants also have been valuable in genetic analysis of the incompatibility locus and as mutagen monitoring systems. The most widely used mutants are those of waxy in barley and maize that affect starch deposition. Genetic analyses of waxy pollen mutants in maize have led to intragenic mapping of the locus and to the development of a genetic system that has already been utilized in mutagen monitoring experiments. Mutants of the waxy locus of barley have been mapped to a lesser extent, and are being developed for mutagen monitoring systems.

From research with the waxy loci of maize and barley, and the *Adh* locus of maize, it appears that most mutants are the result of some type of chromosome changes—probably deletions. As yet, no mutants have proved to be caused by base substitution mutations.

The utilization of pollen mutants for mutagenesis and genetic studies will be greatly enhanced by increased knowledge of the chemistry, structure, and biology of the pollen grain. Furthermore, this activity will progress with more detailed analyses of select pollen mutants, the development of more rapid and accurate techniques for scoring pollen mutants, and adequate statistical techniques for establishing true forward and reverse mutant frequencies and mutation rates.

ACKNOWLEDGMENTS

The contributions of Dr. Alan Hodgdon and Dr. Pablo Arenaz to this paper are gratefully acknowledged. Data and information provided by the authors came from research supported in part by National Institute of Environmental Health Sciences Grant no. ES02224 and Department of Energy Contract no. DE-AM06-76-RL02221. DOE/RL-02221/54.

REFERENCES

Amano, E. 1981a. "Flow System for Automated Analysis of Maize Pollen." *Environ. Health Perspect.* 37:165-168.

——. 1981b. "Genetic and Biochemical Characterization of Waxy Mutants in Cereals." *Environ. Health Perspect.* 37:35-41.

——. 1972. "Genetic Fine Structure Analysis of Mutants Induced by Ethyl Methanesulfonate." *Gamma Field Symp.* 11:43-59.

——. 1968. "Comparison of Ethyl Methanesulfonate and Radiation Induced *Waxy* Mutants in Maize." *Mutation Res.* 5:41-46.

Amano, E., and Harold H. Smith. 1965. "Mutations Induced by Ethyl Methanesulfonate in Maize." *Mutation Res.* 2:344-351.

Ascher, P. D. 1981. "Effects of Exogenous Materials on Pollen Tube Growth in *Lilium longiflorum.*" *Environ. Health Perspect.* 37:107-115.

Barka, T., and P. J. Anderson. 1962. "Histochemical Methods for Acid Phosphatase Using Hexazonium Pararosanilin as Coupler." *J. Histochem. Cytochem.* 10:741-753.

Beckman, L., J. G. Scandalios, and J. L. Brewbaker. 1964. "Genetics of Leucine Aminopeptidase Isozymes in Maize." *Genetics* 50:899-904.

Bianchi, A. 1965. "Some Aspects of Mutagenesis in Maize." In *Induction of Mutations and the Mutation Process*, ed. J. Veleminský and T. Gichner, pp. 30-37. Prague: Publishing House of the Czechoslovak Academy of Sciences.

Bilderback, D. E. 1981. "Impatiens Pollen Germination and Tube Growth as a Bioassay for Toxic Substances." *Environ. Health Perspect.* 37:95-103.

Brewbaker, J. L. 1971. "Pollen Enzymes and Isoenzymes." In *Pollen: Development and Physiology*, ed. J. Heslop-Harrison, pp. 156-170. London: Butterworths.

Brewbaker, J. L., and G. C. Emery. 1962. "Pollen Radiobotany." *Radiat. Bot.* 1:101-154.

Brewbaker, J. L., and A. T. Natarajan. 1960. "Centric Fragments and Pollen-part Mutation of Incompatibility Alleles in *Petunia.*" *Genetics* 45:699-704.

Briggs, R. W., E. Amano, and H. H. Smith. 1965. "Genetic Recombination with Ethyl-methanesulphonate-induced Waxy Mutants in Maize." *Nature* 207:890-891.

Briggs, R. W., and H. H. Smith. 1965. "Effects of X-radiation on Intracistron Recombination at the *Waxy* Locus in Maize." *J. Hered.* 56:157-162.

Brooks, J., and G. Shaw. 1971. "Recent Developments in the Chemistry, Biochemistry, Geochemistry and Post-tetrad Ontogeny of Sporopollenins Derived from Pollen and Spore Exines." In *Pollen: Development and Physiology*, ed. J. Heslop-Harrison, pp. 99-114. London: Butterworths.

Clapham, D. H. 1977. "Haploid Induction in Cereals." In *Applied and Fundamental Aspects of Plant Cell, Tissue and Organ Culture*, ed. J. Reinert and Y. P. S. Bajaj, pp. 279-298. Berlin: Springer Verlag.

Clarke, A. E., and R. B. Knox. 1978. "Cell Recognition in Flowering Plants." *Quart. Rev. Biol.* 53:3-28.

Coe, E. H., and M. G. Neuffer. 1977. "The Genetics of Corn." In *Corn and Corn Improvement*, ed. G. F. Sprague, pp. 111-223. Madison, Wis.: American Society of Agronomy.

Constantin, M. J. 1982. "Plant Genetic Systems with Potential for the Detection

of Atmospheric Mutagens." In *Gene Toxic Effect of Airborn Agents* (in press).
——. 1981. "Pollen Embryogenesis to Induce, Detect, and Analyze Mutants." *Environ. Health Perspect.* 37:27-33.
——. 1978. "Utility of Specific Locus Systems." *Environ. Health Perspect.* 27:69-75.
Darlington, C. D., and L. F. La Cour. 1969. *The Handling of Chromosomes.* 5th ed. London: George Allen and Unwin.
Davies, D. R., and D. A. Hopwood, eds. 1980. *The Plant Genome.* Norwich, England: The John Innes Charity, John Innes Institute.
Devreux, M., and D. de Nettancourt. 1974. "Screening Mutations in Haploid Plants." In *Haploids in Higher Plants*, ed. K. J. Kasha, pp. 309-322. Guelph, Canada: University of Guelph.
Dickinson, H. G., and D. Lewis. 1973. "The Formation of the Tryphine Coating the Pollen Grains of *Raphanus* and Its Properties Relating to the Self-incompatibility System." *Proc. R. Soc. Lond.* B 184:149-165.
Efron, Y. 1973. "Inheritance Studies with Inbred Lines of Maize Having Different Activity Levels of the AP_1 Controlled Acid Phosphatase Isozymes." *Theoret. Appl. Genet.* 43:323-328.
——. 1971. "Differences Between Maize Inbreds in the Activity Level of the AP_1-Controlled Acid Phosphatase." *Biochem. Genet.* 5:33-44.
——. 1970. "Tissue Specific Variation in the Isozyme Pattern of the AP_1 Acid Phosphatase in Maize." *Genetics* 65:575-583.
——. 1969. "Specific Inhibition of Acid Phosphatase-3 in Pollen of Maize by the Diazonium Salt Fast Garnet GBC." *J. Histochem. Cytochem.* 17:734-737.
Ehrenberg, L. 1977. "Aspects of Statistical Inference in Testing for Genetic Toxicity." In *Handbook of Mutagenicity Test Procedures*, ed. B. J. Kilbey, M. Legator, W. Nichols, and C. Ramel, pp. 419-459. Amsterdam: Elsevier.
——. 1971. "Higher Plants." In *Chemical Mutagens. Principles and Methods for Their Detection*, ed. A. Hollaender, pp. 365-386. New York: Plenum.
Ehrenberg, L., and G. Eriksson. 1966. "The Dose Dependence of Mutation Rates in the Rad Range, in the Light of Experiments with Higher Plants." *Acta Radiol. Suppl.* 254:73-81.
Erdtman, G. 1952. *Pollen Morphology and Plant Taxonomy.* Uppsala: Almqvist and Wiksells.
——. 1945. "Pollen Morphology and Taxonomy. V. On the Occurrence of Tetrads and Dyads." *Svensk. Bot. Tidskr.* 39:286-297.
Eriksson, G. 1971. "Variation in Radiosensitivity and the Dose Effect Relationship in the Low Dose Region." *Hereditas* 68:101-114.
——. 1969. "The Waxy Character." *Hereditas* 63:180-204.
——. 1963. "Induction of Waxy Mutants in Maize by Acute and Chronic Gamma Irradiation." *Hereditas* 50:161-178.
——. 1962. "Radiation Induced Reversions of a Waxy Allele in Barley." *Radiat. Bot.* 2:35-39.

Faure, F., M. Delpoux, and M. A. Dalebroux. 1981. "Genetic Effects of Low Chronic Doses of γ-irradiation on the *Waxy* Locus of Barley." *Mutation Res.* 81:59-62.

Feder, W. A. 1981. "Bioassaying for Ozone with Pollen Systems." *Environ. Health Perspect.* 37:117-123.

Ferl, R. J., M. D. Brennan, and D. Schwartz. 1980. "*In Vitro* Translation of Maize ADH: Evidence for the Anaerobic Induction of mRNA." *Biochem. Genet.* 18:681-691.

Freeling, M. 1981. "Toward Monitoring Specific DNA Lesions in the Gene by Using Pollen Systems." *Environ. Health Perspect.* 37:13-17.

——. 1978a. "Allelic Variation at the Level of Intragenic Recombination." *Genetics* 89:211-224.

——. 1978b. "Maize Adh1 as a Monitor of Environmental Mutagens." *Environ. Health Perspect.* 27:91-97.

——. 1977. "Spontaneous Forward Mutation Versus Reversion Frequencies for Maize *Adh*1 in Pollen." *Nature* 267:154-156.

——. 1976. "Intragenic Recombination in Maize: Pollen Analysis Methods and the Effect of Parental *Adh*1$^+$ Isoalleles." *Genetics* 83:701-717.

Freeling, M., and D. S. K. Cheng. 1978. "Radiation-induced Alcohol Dehydrogenase Mutants in Maize Following Allyl Alcohol Selection of Pollen." *Genet. Res.* 31:107-129.

Freeling, M., and D. Schwartz. 1973. "Genetic Relationships Between Multiple Alcohol Dehydrogenases of Maize." *Biochem. Genet.* 8:27-36.

Freeling, M., and J. C. Woodman. 1979. "Regulatory Variant and Mutant Alleles in Higher Organisms and Their Possible Origin via Chromosomal Breaks." In *The Plant Seed*, ed. I. Rubenstein, R. Phillips, C. Green, and B. G. Gengenbach, pp. 85-111. New York: Academic Press.

Frydenberg, O., and G. Nielsen. 1966. "Amylase Isozymes in Germinating Barley Seeds." *Hereditas* 54:123-139.

Frydenberg, O., G. Nielsen, and J. Sandfaer. 1969. "The Inheritance and Distribution of α-amylase Types and DDT Responses in Barley." *Z. Pflanzenzüchtg.* 61:201-215.

Gentile, J. M., E. D. Wagner, and M. J. Plewa. 1978. "A Comprehensive Analysis of the Mutagenic Properties of Pesticides Used in Commercial Maize Production." *Mutation Res.* 53:112 (abs.).

Heslop-Harrison, J. 1975a. "The Physiology of the Pollen Grain Surface." *Proc. R. Soc. Lond.* B 190:275-299.

——. 1975b. "Incompatibility and the Pollen-Stigma Interaction." *Ann. Rev. Plant Physiol.* 26:403-425.

——. 1971. "The Pollen Wall: Structure and Development." In *Pollen: Development and Physiology*, ed. J. Heslop-Harrison, pp. 75-98. London: Butterworths.

Heslop-Harrison, J., and Y. Heslop-Harrison. 1973. "Pollen-wall Proteins: 'Gametophytic' and 'Sporophytic' Fractions in the Pollen Walls of the Malvaceae." *Ann. Bot.* 37:403-412.

——. 1970. "Evaluation of Pollen Viability by Enzymically Induced Fluorescence; Intracellular Hydrolysis of Fluorescein Diacetate." *Stain Tech.* 45:115-120.

Heslop-Harrison, J., R. B. Knox, and Y. Heslop-Harrison. 1974. "Pollen-wall Proteins: Exine-held Fractions Associated with the Incompatibility Response in Cruciferae." *Theoret. Appl. Genet.* 44:133-137.

Heslop-Harrison, J., R. B. Knox, Y. Heslop-Harrison, and O. Mattsson. 1975. "Pollen-wall Proteins: Emission and Role in Incompatibility Responses." In *The Biology of the Male Gamete*, ed. J. G. Duckett and P. A. Racey, pp. 189-202. New York: Academic Press.

Hockett, E. A., and R. F. Eslick. 1971. "Genetic Male-Sterile Genes Useful in Hybrid Barley Production." In *Barley Genetics II*, ed. R. A. Nilan, pp. 298-307. Pullman: Washington State University Press.

——. 1968. "Genetic Male Sterility in Barley. I. Nonallelic Genes." *Crop Sci.* 8:218-220.

Hodgdon, A. L., A. H. Marcus, P. Arenaz, J. L. Rosichan, T. P. Bogyo, and R. A. Nilan. 1981. "Ontogeny of the Barley Plant as Related to Mutation Expression and Detection of Pollen Mutations." *Environ. Health Perspect.* 37:5-7.

Howlett, B. J., R. B. Knox, and J. Heslop-Harrison. 1973. "Pollen-Wall Proteins: Release of the Allergen Antigen E from Intine and Exine Sites in Pollen Grains of Ragweed and *Cosmos*." *J. Cell Sci.* 13:603-619.

International Atomic Energy Agency. 1977. *Manual on Mutation Breeding.* Technical Reports Series no. 119. 2nd ed. Vienna: International Atomic Energy Agency.

Jackson, J. F., and H. F. Linskens. 1980. "DNA Repair in Pollen: Range of Mutagens Inducing Repair, Effect of Replication Inhibitors and Changes in Thymidine Nucleotide Metabolism During Repair." *Molec. Gen. Genet.* 180:517-522.

——. 1979. "Pollen DNA Repair After Treatment with the Mutagens 4-Nitroquinoline-1-oxide, Ultraviolet and Near-Ultraviolet Irradiation, and Boron Dependence of Repair." *Molec. Gen. Genet.* 176:11-16.

——. 1978. "Evidence for DNA Repair After Ultraviolet Irradiation of *Petunia hybrida* Pollen." *Molec. Gen. Genet.* 161:117-120.

Jones, R. L., and R. F. Chen. 1976. "Immunohistochemical Localization of α-amylase in Barley Aleurone Cells." *J. Cell Sci.* 20:183-198.

Katz, A. J. 1979. "Design and Analysis of Experiments on Mutagenicity. II. Assays Involving Microorganisms." *Mutation Res.* 64:61-77.

——. 1978. "Design and Analysis of Experiments on Mutagenicity. I. Minimal Sample Sizes." *Mutation Res.* 50:301-307.

King, T. P., P. S. Norman, and J. T. Connell. 1964. "Isolation and Characterization of Allergens from Ragweed Pollen. II." *Biochemistry* 3:458-468.

Knox, R. B. 1981. "Cell Surface Phenomena: Pollen-Pistil Interactions." In *Encyclopedia of Plant Physiology*, new ser., ed. W. Tanner and F. A. Loewus (forthcoming).

——. 1971. "Pollen-wall Protein: Localization, Enzymic and Antigenic Activity During Development in *Gladiolus* (Iridaceae)." *J. Cell Sci.* 9:209-237.

Knox, R. B., and J. Heslop-Harrison. 1971. "Pollen-Wall Proteins: Localization of Antigenic and Allergenic Proteins in the Pollen-Grain Walls of *Ambrosia* spp. (Ragweeds)." *Cytobios* 4:49-54.

——. 1970. "Pollen-Wall Proteins: Localization and Enzymic Activity." *J. Cell Sci.* 6:1-27.

Knox, R. B., J. Heslop-Harrison, and Y. Heslop-Harrison. 1975. "Pollen-Wall Proteins." In *The Biology of the Male Gamete*, ed. J. G. Duckett and P. A. Racey, pp. 177-187. London: Academic Press.

Knox, R. B., R. R. Willing, and A. E. Ashford. 1972. "Role of Pollen-Wall Proteins as Recognition Substances in Interspecific Incompatibility in Poplars." *Nature* (London) 237:381-383.

Konzak, C. F., and W. R. Singleton. 1956a. "The Effects of Thermal-Neutron Radiation on Mutation of Endosperm Loci in Maize." *Proc. Nat. Acad. Sci. U.S.A.* 42:78-84.

——. 1956b. "The Mutation of Linked Maize Endosperm Loci Induced by Thermal-Neutron, X-, Gamma, and Ultraviolet Radiation." *Proc. Nat. Acad. Sci. U.S.A.* 42:239-245.

Laine, S. 1977. "Diffusion of Proteins by Intact Birch Pollen Grains: Enzymic and Antigenic Activity." *Grana* 16:187-189.

Larsen, K. 1977. "Self Incompatibility in *Beta vulgaris* L. I. Four Gametophytic, Complementary S-loci in Sugar Beet." *Hereditas* 85:227-248.

Lee, Y. S., and D. B. Dickinson. 1979. "Characterization of Pollen Antigens from *Ambrosia* L. (Compositae) and Related Taxa by Immunoelectrophoresis and Radial Immunodiffusion." *Amer. J. Bot.* 66:245-252.

Lewis, D. 1960. "Genetic Control of Specificity and Activity of the S Antigen in Plants." *Proc. R. Soc. Lond.* B 151:468-471.

——. 1949. "Structure of the Incompatibility Gene. II. Induced Mutation Rate." *Heredity* 3:339-355.

——. 1947. "Competition and Dominance of Incompatibility Alleles in Diploid Pollen." *Heredity* 1:85-108.

Lewis, D., S. Burrage, and D. Walls. 1967. "Immunological Reactions of Single Pollen Grains, Electrophoresis and Enzymology of Pollen Protein Exudates." *J. Exp. Bot.* 18:371-378.

Li, S. L., and G. P. Rédei. 1969. "Estimation of Mutation Rate in Autogamous Diploids." *Radiat. Bot.* 9:125-131.

Lindgren, D. 1975. "Sensitivity of Premeiotic and Meiotic Stages to Spontaneous and Induced Mutations in Barley and Maize." *Hereditas* 79:227-238.

Lindgren, D., and G. Eriksson. 1971. "The Mutated Sector in Barley Spikes Following Isopropyl Methanesulfonate (iPMS) Treatment." *Hereditas* 69:129-133.

Lindgren, D., and K. Lindgren. 1972. "Investigations of Environmental Mutagens by the Waxy Method." *EMS Newsletter* 6:22 (abs.).

Lower, W. B., P. S. Rose, and V. K. Drobney. 1978. "In Situ Mutagenic and Other Effects Associated with Lead Smelting." *Mutation Res.* 54:83-93.

Lundquist, A. 1975. "Complex Self-Incompatibility Systems in Angiosperms." *Proc. R. Soc. Lond.* B 188:235-245.

——. 1954. "Studies on Self-Sterility in Rye; *Secale cereale* L." *Hereditas* 40:278-294.

Ma, T.-H. 1981. "*Tradescantia* Micronucleus Bioassay and Pollen Tube Chromatid Aberration Test for *in Situ* Monitoring and Mutagen Screening." *Environ. Health Perspect.* 37:85-90.

——. 1979. "Micronuclei Induced by X-rays and Chemical Mutagens in Meiotic Pollen Mother Cells of *Tradescantia.*" *Mutation Res.* 64:307-313.

Ma, T.-H., A. H. Sparrow, L. A. Schairer, and A. F. Nauman. 1978. "Effect of 1,2-Dibromoethane (DBE) on Meiotic Chromosomes of *Tradescantia.*" *Mutation Res.* 58:251-258.

Maheshwari, P. 1949. "The Male Gametophyte of Angiosperms." *Bot. Rev.* 15:1-75.

Mäkinen, Y., and T. Macdonald. 1968. "Isoenzyme Polymorphism in Flowering Plants. II. Pollen Enzymes and Isoenzymes." *Physiol. Plant.* 21:477-486.

Marcus, A., and S. Czajkowski. 1979. "First Passage Times as Environmental Safety Indicators; Carboxyhemoglobin from Cigarette Smoke." *Biometrics* 35:539-548.

Mascarenhas, J. P. 1981. "Microspore and Microgametophyte Development in Relation to Biological Activity of Environmental Pollutants." *Environ. Health Perspect.* 37:9-12.

——. 1975. "The Biochemistry of Angiosperm Pollen Development." *Bot. Rev.* 41:259-313.

Melamed, M. R., L. A. Kamentsky, and E. A. Boyse. 1969. "Cytotoxic Test Automation: A Live-Dead Cell Differential Counter." *Science* 163: 285-286.

Moore, C. W., and R. G. Creech. 1972. "Genetic Fine Structure Analysis of the *amylose-extender* Locus in *Zea mays* L." *Genetics* 70:611-619.

Mulcahy, D. L. 1981. "Pollen Tetrads in the Detection of Environmental Mutagenesis." *Environ. Health Perspect.* 37:91-94.

Mulcahy, D. L., and C. M. Johnson. 1978. "Self-Incompatibility Systems as Bioassays for Mutagens." *Environ. Health Perspect.* 27:85-90.

Nelson, O. E. 1976. "Previously Unreported *wx* Heteroalleles." *Maize Genet. Coop. News Letter* 50:109-113.

——. 1975. "The *Waxy* Locus in Maize. III. Effect of Structural Heterozygosity on Intragenic Recombination and Flanking Marker Assortment." *Genetics* 79:31-44.

——. 1968. "The *Waxy* Locus in Maize. II. The Location of the Controlling Element Alleles." *Genetics* 60:507-524.

——. 1962. "The *Waxy* Locus in Maize. I. Intralocus Recombination Frequency Estimates by Pollen and by Conventional Analyses." *Genetics* 47: 737-742.

——. 1959. "Intracistron Recombination in the *Wx/wx* Region in Maize." *Science* 130:794-795.

Nelson, O. E., and B. Burr. 1973. "Biochemical Genetics of Higher Plants." *Ann. Rev. Plant Physiol.* 24:493-518.

Nelson, O. E., P. S. Chourey, and M. T. Chang. 1978. "Nucleoside Diphosphate Sugar-Starch Glucosyl Transferase Activity of *wx* Starch Granules." *Plant Physiol.* 62:383-386.

Nelson, O. E., and H. W. Rines. 1962. "The Enzymatic Deficiency in the Waxy Mutant of Maize." *Biochem. Biophys. Res. Commun.* 9:297-300.

Nelson, O. E., and C. Y. Tsai. 1964. "Glucose Transfer from Adenosine Diphosphate-glucose to Starch in Preparations of Waxy Seeds." *Science* 145: 1194-1195.

Nettancourt, D. de, M. Devreux, F. Carluccio, U. Laneri, M. Cresti, E. Pacini, G. Sarfatti, and A. J. G. van Gastel. 1975. "Facts and Hypotheses on the Origin of S Mutations and on the Function of the S Gene in *Nicotiana alata* and *Lycopersicum peruvianum.*" *Proc. R. Soc. Lond.* B 188: 345-360.

Nettancourt, D. de, G. Eriksson, D. Lindgren, and K. Puite. 1977. "Effects of Low Doses by Different Types of Radiation on the Waxy Locus in Barley and Maize." *Hereditas* 85:98-100.

Neuffer, M. G., and E. H. Coe, Jr. 1974. "Corn (Maize)." In *Handbook of Genetics.* Vol. 2, *Plants, Plant Viruses, and Protists*, ed. R. C. King, pp. 3-30. New York: Plenum.

Neuffer, M. G., and G. Ficsor. 1963. "Mutagenic Action of Ethyl Methanesulfonate in Maize." *Science* 139:1296-1297.

Nilan, R. A. 1978. "Potential of Plant Genetic Systems for Monitoring and Screening Mutagens." *Environ. Health Perspect.* 27:181-196.

——. 1967. "Nature of Induced Mutations in Higher Plants." In *Induced Mutations and Their Utilization, Erwin-Bauer Memorial Lectures IV*, ed. K. Gröber, F. Scholz, and M. Zacharias, pp. 5-20. Berlin: Akademie Verlag.

——. 1964. *The Cytology and Genetics of Barley, 1951-1962.* Pullman: Washington State University Press.

Nilan, R. A., A. Kleinhofs, and R. L. Warner. 1981. "Use of Induced Mutants of Genes Controlling Nitrate Reductase, Starch Deposition, and Anthocyanin Synthesis in Barley." In *International Symposium on Induced Mutations as a Tool for Crop Plant Improvement.* Vienna: International Atomic Energy Agency (in press).

Nilan, R. A., J. L. Rosichan, P. Arenaz, A. L. Hodgdon, and A. Kleinhofs. 1981. "Pollen Genetic Markers for Detection of Mutagens in the Environment." *Environ. Health Perspect.* 37:19-25.

Nilan, R. A., and B. K. Vig. 1976. "Plant Test Systems for Detection of Chemical Mutagens." In *Chemical Mutagens. Principles and Methods for Their Detection*, ed. A. Hollaender, pp. 143-170. New York: Plenum.

Østerbye, U. 1975. "Self-Incompatibility in *Ranunculus acris* L." *Hereditas* 80: 91-112.

Owais, W. M., M. A. Zarowitz, R. A. Gunovich, A. L. Hodgdon, A. Kleinhofs, and R. A. Nilan. 1978. "A Mutagenic in Vivo Metabolite of Sodium Azide." *Mutation Res.* 53:355-358.

Pandey, K. 1967. "Elements of the S-gene Complex. II. Mutation and Complementation at the S_1 Locus of *Nicotiana alata*." *Heredity* 22: 255-283.
Pfahler, P. L. 1981. "*In Vitro* Germination Characteristics of Maize Pollen to Detect Biological Activity of Environmental Pollutants." *Environ. Health Perspect.* 37:125-132.
——. 1975. "Factors Affecting Male Transmission in Maize (*Zea mays* L.)." In *Gamete Competition in Plants and Animals*, ed. D. L. Mulcahy, pp. 115-124. Amsterdam: Elsevier.
Pinkel, D. 1981. "On the Possibility of Automated Scoring of Pollen Mutants." *Environ. Health Perspect.* 37:133-136.
Plewa, M. J. 1982. "Specific Locus Mutation Assays in *Zea mays*." *Gene Tox Report for EPA.* (forthcoming).
——. 1981. "Specific Locus Assays in *Zea mays*." *Mutation Res.* (in press).
——. 1978. "Activation of Chemicals into Mutagens by Green Plants: A Preliminary Discussion." *Environ. Health Perspect.* 27:45-50.
Plewa, M. J., and J. M. Gentile. 1981. "The Production of Mutagens by Plant Systems." In *Chemical Mutagens. Principles and Methods for Their Detection*, Vol. 7, eds. F. J. de Serres and A. Hollaender. New York: Plenum (in press).
——. 1976. "Mutagenicity of Atrazine: A Maize-Microbe Bioassay." *Mutation Res.* 38:287-292.
Plewa, M. J., and E. D. Wagner. 1981. "Germinal Cell Mutagenesis in Specially Designed Maize Genotypes." *Environ. Health Perspect.* 37:61-73.
Porter, E. K. 1981. "Origins and Genetic Nonvariability of the Proteins Which Diffuse from Maize Pollen." *Environ. Health Perspect.* 37:53-59.
Preiss, J., and C. Levi. 1980. "Starch Biosynthesis and Degradation." In *The Biochemistry of Plants*, ed. P. K. Stumpf and E. E. Conn, Vol. 3, pp. 371-423. New York: Academic Press.
Rick, C. M. 1974. "The Tomato." In *Handbook of Genetics.* Vol. 2, *Plants, Plant Viruses, and Protists*, ed. R. C. King, pp. 247-280. New York: Plenum.
Rosen, W. G. 1981. "Pollen Systems to Detect Phytotoxicants in the Environment: An Introduction." *Environ. Health Perspect.* 37:105.
Rosichan, J. L., P. Arenaz, N. Blake, A. Hodgdon, A. Kleinhofs, and R. A. Nilan. 1981. "An Improved Method for the Detection of Mutants at the *Waxy* Locus in *Hordeum vulgare*." *Environ. Mutagenesis* 3:91-93.
Schairer, L. A., J. van't Hof, C. G. Hayes, R. M. Burton, and F. J. de Serres. 1978. "Exploratory Monitoring of Air Pollutants for Mutagenicity Activity with the *Tradescantia* Stamen Hair System." *Environ. Health Perspect.* 27:51-60.
Schmid, W. 1975. "The Micronucleus Test." *Mutation Res.* 31:9-15.
Schwartz, D. 1981. "*Adh* Locus in Maize for Detection of Mutagens in the Environment." *Environ. Health Perspect.* 37:75-77.
——. 1975. "The Molecular Basis for Allelic Complementation of Alcohol Dehydrogenase Mutants of Maize." *Genetics* 79:207-212.
——. 1971a. "Dimerization Mutants of Alcohol Dehydrogenase of Maize." *Proc. Nat. Acad. Sci. U.S.A.* 68:145-146.

——. 1971b. "Genetic Control of Alcohol Dehydrogenase—a Competition Model for Regulation of Gene Action." *Genetics* 67:411-425.

Schwartz, D., and J. Osterman. 1976. "A Pollen Selection System for Alcohol-Dehydrogenase-Negative Mutants in Plants." *Genetics* 83:63-65.

Scott, B. R., A. H. Sparrow, S. S. Schwemmer, and L. A. Schairer. 1978. "Plant Metabolic Activation of 1,2-Dibromoethane (EDB) to a Mutagen of Greater Potency." *Mutation Res.* 49:203-212.

de Serres, F. J. 1978. "Introduction: Utilization of Higher Plant Systems as Monitors of Environmental Mutagens." *Environ. Health Perspect.* 27:3-6.

Sree Ramulu, K., H. Schibilla, and P. Dijkhuis. 1981. "Self-Incompatibility System of *Oenothera organensis* for the Detection of Genetic Effects at Low Radiation Doses." *Environ. Health Perspect.* 37:43-51.

Stadler, L. J. 1928. "Genetic Effects of X-rays in Maize." *Proc. Nat. Acad. Sci. U.S.A.* 14:69-75.

Sulovská, K., D. Lindgren, G. Eriksson, and L. Ehrenberg. 1969. "The Mutagenic Effect of Low Concentrations of Ethylene Oxide in Air." *Hereditas* 62:264-266.

Tates, A. D., I. Neuteboom, M. Hofker, and L. den Engelse. 1980. "A Micronucleus Technique for Detecting Clastogenic Effects of Mutagens/Carcinogens (DEN, DMN) in Hepatocytes of Rat Liver in Vivo." *Mutation Res.* 74:11-20.

Tsai, C. Y. 1974. "The Function of the Waxy Locus in Starch Biosynthesis in Maize Endosperm." *Biochem. Genet.* 11:83-96.

Tyrer, H. W. 1981. "Technology for Automated Analysis of Maize Pollen Used as a Marker for Mutation: 1. Flow-Through Systems." *Environ. Health Perspect.* 37:137-142.

Vithanage, H. I. M. V., and R. B. Knox. 1979. "Pollen Development and Quantitative Cytochemistry of Exine and Intine Enzymes in Sunflower, *Helianthus annuus* L." *Ann. Bot.* 44:95-106.

Watson, L., and R. B. Knox. 1976. "Pollen Wall Antigens and Allergens: Taxonomically-Ordered Variation Among Grasses." *Ann. Bot.* 40:399-408.

Weber, D. F. 1981. "Maize Pollen Test Systems to Detect Nondisjunction." *Environ. Health Perspect.* 37:79-84.

Weeke, B., and H. Lowenstein. 1973. "Allergens Identified in Crossed Radioimmunoelectrophoresis." *Scand. J. Immun.* 2 (suppl. 1):149-154.

Zimmermann, F. K., F. J. de Serres, M. D. Shelby, and J. S. Wassom. 1979. "Systems to Detect Induction of Aneuploidy by Environmental Mutagens." *Environ. Health Perspect.* 31:1-167.

4

Using Components of the Native Flora to Screen Environments for Mutagenic Pollutants

Edward J. Klekowski, Jr.

INTRODUCTION

The screening of environments for dispersed mutagens usually is based upon two assay protocols: (1) a sample of the environment (gas, liquid, solid, or even living phase; see Chapter 5 in this volume) is concentrated, extracted, and tested with a diversity of laboratory-based mutagen assays, or (2) in situ mutagen assays are employed that monitor the environment directly and detect ambient mutagen concentrations (see Chapter 6 in this volume). Although numerous theoretical advantages can be postulated in favor of in situ mutagen assays (Lower, Rose, and Drobney, 1978), the use of such assays is uncommon. Most environmental monitoring is conducted by means of chemical concentration and extraction procedures coupled with laboratory-based mutagen assays because such assays (especially microbial assays) are well characterized genetically, give repeatable results, work in a short time frame, can be adapted for metabolic promutagen activation, and allow the researcher to study fractions of the environmental sample in order to chemically identify the mutagen(s).

In spite of the great utility of microbial mutagen assays, environmental monitoring based entirely upon such assay protocols can result in numerous false negatives. Screening polluted environments for dispersed mutagens presents special problems; very often pollutants in such environments may be nonrandomly distributed (reflecting changes in air and wind currents) and discharged episodically. In addition, the mutagens may be in low concentration and thus require large environmental samples in order to be detected with microbial assays. Johnston and Hopke (1980) studied the sensitivity of the Ames test in terms of screening water samples for dispersed mutagens. Using the published dose-response data for various mutagens, they calculated the dose necessary to double the revertant count in an Ames test and determined a distribution of doubling doses for these mutagens. The sample volume necessary to encompass the doubling doses of 95 percent of all mutagens, assuming 1 μg of mutagen per liter, is 1,500 liters of water. Under actual conditions Johnston and Hopke (1980) conclude that this is probably an optimistic estimate and that the sample volume should be larger.

Three critical problems are thus involved in sampling environments for mutagens: (1) nonrandom distribution in the environment, (2) episodic distribution in terms of both mutagen discharge and movement, and (3) low concentration. In situ mutagen assays based upon the measurement of mutations in the native flora can overcome these problems. If perennials are selected as bioassays, these assay organisms have been in place and exposed for a long time, and the various routes of mutagen entry into these plants insure a total environmental sample. Plants take up nutrients and water from the soil, and also sample the gaseous atmosphere through stomata. Thus the long exposure can overcome the low mutagen concentration, and the presence of the bioassay organism in the environment for a long time (years) will increase the probability of detecting episodes of mutagen pollution. The primary drawback to such bioassays is the unknown capacity of plant metabolism in the activation of promutagens and how this metabolism compares with hepatic activation in mammals (see Chapters 1 and 2 in Volume I).

The present paper will develop some general principles that must be considered in the development of long-term in situ mutagen assays. The relative probabilities of mutation induction, fixation, and detection will be considered with reference to the anatomy, morphology, and apical ontogeny of vascular plants. The development and use of plant mutagen assays must be based upon an understanding of the growth and development of these organisms under conditions of long-term, chronic exposure to mutagens. In addition to the general relations of mutagenesis and vascular plant biology, the use of ferns as mutagen assays will be described in depth.

PLANT ANATOMY AND MUTAGENESIS

The measurement of postzygotic mutational damage in vascular plants requires both an awareness of the limitations of such measurements and a knowledge of the biology of these organisms in terms of mutation induction and fixation. Since in situ assays are based upon the native biota, experiments have not been conducted that document the true genetic nature of the aberrant phenotypes being screened. This problem can be resolved, at least in part, by collecting the data in such a fashion that Mendelian ratios can be detected. Thus the genetic basis of these aberrant phenotypes could be implied at least by their association with 1:1 ratios of meiotic products or Mendelian zygotic ratios. The collection of data so that genetic ratios are preserved requires an understanding of the dynamics of mutation induction and fixation in the somatic tissues of higher plants.

Basic to mutation induction in both seeds and adult plants is the ontogeny of the shoot apex. The shoot apex, with its constituent apical meristems, is responsible for the primary development of the shoot, including the reproductive organs. The plant may consist of a single shoot apex (an embryo within a seed) or a multitude of shoot apices (a tree with its branches and axillary buds). Comparative studies of various groups of vascular plants have shown a diverse array of structural patterns (Steeves and Sussex, 1972).

In angiosperms, typically a tunica-corpus apical anatomy is present. The apical dome shows a definite stratification of the superficial cell layers. The two outermost layers above the youngest leaf primordia are characterized by cell divisions that are at right angles to the surface of the dome; these are the tunica layers. The corpus is within these layers, and is characterized by planes of cell division that appear to be random. In many gymnosperms apical meristems have been described that exhibit apical zonation. Thus, rather than distinct layers, the apices show cytohistological zonation with cytologically recognizable groups of cells. These two types of apical organization show numerous qualitative and quantitative variations in nature, and should only be considered idealized types (see Steeves and Sussex [1972] for further discussion).

In pteridophytes, ferns and fern allies, a third kind of apical organization and ontogeny occurs. In these plants the apical meristem is based upon a single four-sided, pyramid-shaped apical cell. This cell has three cutting faces in terms of division planes, and the products of these divisions result in lineages of cells that also divide and ultimately give rise to the bulk of the fern rhizome (stem) as well as the fronds (leaves, both vegetative and reproductive) (Bierhorst, 1977; Gifford, Polito, and Nitayangkura, 1979). In contrasting the seed plant (angiosperm and gymnosperm) shoot apex with the pteridophyte shoot apex in terms of mutation induction and storage, the critical difference is in the number of apical initials. In pteridophytes there is a single, stable apical initial, the apical cell, whereas seed plants have a number of apical initials per shoot apex. Because of these differences in organization and number of initials, stable mericlinal, periclinal, and sectorial chimeras are impossible in pteridophytes. This is not the case for seed plants, in which such phenomena have often been documented (Stewart, 1978).

To appreciate the associated problems in detecting mutations in vascular plants, it is necessary to consider the shoot apices of these plants in terms of mutation induction and fixation. Fixation is either the development of stable chimeras in the shoot apex that give rise to reproductive tissue, or the transformation of the entire apex into the mutant genotype. The subsequent discussion will mathematically compare the probabilities of mutation induction and fixation in pteridophyte (fern) and seed plant apical meristems. The outcome of these formulations will show the kinds of sampling strategies necessary in screening components of the native flora for postzygotic mutational damage.

In the following analysis, the probability of a mutation occurring equals A, the probability of a mutation not occurring equals B, and $A + B = 1$. The number of target cells (apical initials) that have some probability of fixing the mutation is given by n; thus, in ferns with a single apical cell per apex, $n = 1$, whereas in seed plants $n > 1$. Thus the probability of one or more target cells being mutated is found by expanding the binomial

$$(A + B)^n = \sum_{k=0}^{n} C_{n\,k}\, A^{n-k}B^k = \sum_{k=0}^{n} \frac{n!}{(n-k)!k!} A^{n-k}B^k \qquad (1)$$

and summing all terms of this expansion, excluding the last term, B^n; thus the probability that one or more target cells have mutated is

$$1-B^n \qquad (2)$$

Since the value of A is very small, the value of B is very close to 1. Thus, in the summation (1) the next-to-last term, in which $k = n-1$, is the only value with any magnitude. Therefore

$$\frac{n!}{[n-(n-1)]![n-1]!} A^{n-(n-1)}B^{n-1} \simeq 1-B^n \qquad (3)$$

$$nA^{n-(n-1)}B^{n-1} \simeq 1-B^n \qquad (4)$$

since $B \simeq 1$

$$nA \simeq 1-B^n \qquad (5)$$

Thus the probability that one or more target cells have mutated is closely approximated by nA.

The frequency of shoot apices having at least one mutant initial is nA. In order to determine the sample size necessary to include such a shoot apex, the question must be phrased in terms of the assurity of success. If s is the sample size, the total distribution of the various combinations of shoot apices with and without mutant initials is given by

$$[nA + (1-nA)]^s = 1 \tag{6}$$

Assuming an assurity of 95 percent, the necessary sample size (s) is given by

$$[nA + (1-nA)]^s - (1-nA)^s = .95 \tag{7}$$

$$1-(1-nA)^s = .95 \tag{8}$$

$$(1-nA)^s = 1-.95 \tag{9}$$

$$(1-nA)^s = .05 \tag{10}$$

$$s \log(1-nA) = \log .05 \tag{11}$$

$$s = \frac{\log .05}{\log(1-nA)} \tag{12}$$

The value s is the sample size necessary to include a shoot apex with at least one mutant initial. This is not the sample size that will include a detectable mutant shoot apex. In order to appreciate this distinction, the fate of a mutant apical initial during apical ontogeny must be considered. For example, in the case of a shoot apex with two apical initials where one initial is heterozygous for a mutation, the shoot apex initially has half of the initial mutant (a) and half nonmutant (b). These initials divide mitotically, and from this population of daughter cells a new set of initials is selected at random. The various combinations of mutant and nonmutant initials are given by

$$(a + b)^2 = a^2 + 2ab + b^2 \tag{13}$$

where a^2 is the frequency of apices with both initials heterozygous for the mutation ($a^2 = [1/2]^2 = 1/4$) and b^2 is the frequency of apices where both apical initials lack the mutation (1/4). The frequency of apices that are a chimera is 2ab, or 1/2. Since the frequency of apical initials that are mutant-free still equals the frequency of apical initials that are heterozygous for the mutation in these apices, the initial process is repeated in terms of cell divisions and selection of new initials. Thus the frequency of shoot apices that are chimeras, have lost, or have fixed the mutation is given by

$$1/2\ (a^2 + 2ab + b^2) \tag{14}$$

Therefore, with the second cycle of division and random selection of apical initials, the frequency of shoot apices that have fixed or lost the mutation is increased. The frequency of shoot apices that have fixed the mutation is

$$a^2 + 1/2a^2 = 3/8 \tag{15}$$

Similarly, the frequency of shoot apices that have lost the mutation is

$$b^2 + 1/2\ b^2 = 3/8 \tag{16}$$

and the frequency of shoot apices that are still a chimera is

$$1/2(2ab) = 2/8 \tag{17}$$

The shoot apices that retain the mutant and nonmutant initials (chimeras) will repeat the above processes in the next cycle of cell divisions and apical initial selection. Table 4.1 shows the results of such cycles of cell division and apical initial selection. It is obvious that the frequency of shoot apices that have fixed the mutation asymptotically approaches 1/2 or 1/n, where n is the number of apical initials.

In the case where the mean number of apical initials is three,

TABLE 4.1
Fixation and Loss of Mutant Initials in Shoot Apices Where Two Apical Initials Are Selected Randomly after Each Cycle of Division

	Frequency of Shoot Apices		
Cycles of Cell Division and Selection of Apical Initials	*Chimera*	*Mutant Fixed*	*Mutant Lost*
1	2/4	1/4	1/4
2	2/8	3/8	3/8
3	2/16	7/16	7/16
4	2/32	15/32	15/32

Note: The starting condition consisted of a shoot apex with one apical initial heterozygous for the mutation and one homozygous wild type.

of which one is heterozygous for a mutation (a), a similar pattern of mutation fixation occurs. Thus $a = 1/3$ and $b = 2/3$, and after one cycle of division and random selection of apical initials (3), the frequency of the various types of apices is

$$(a + b)^3 = a^3 + 3a^2b + 3ab^2 + b^3 \tag{18}$$

where the frequency of apices with three initials heterozygous is

$$a^3 = (1/3)^3 = 1/27 \tag{19}$$

the frequency of apices that lack the mutation is

$$b^3 = (2/3)^3 = 8/27 \tag{20}$$

and the frequency of apices that are still a chimera is

$$3a^2b + 3ab^2 = 18/27 \tag{21}$$

In the next cycle of division and selection of initials, $3a^2b$ and $3ab^2$ again yield apices that have either fixed or lost the mutation. Apices described by the term $3a^2b$ have two initials that carry the mutant and one that does not. Thus $a = 2/3$ and $b = 1/3$, and from (18) the frequency of shoot apex types after the second cycle of division and selection is

$$6/27(a^3 + 3a^2b + 3ab^2 + b^3) \tag{22}$$

$$a^3 = 6/27(2/3)^3 \tag{23}$$

$$3a^2b = 6/27(3\cdot(2/3)^2(1/3)) \tag{24}$$

$$3ab^2 = 6/27(3\cdot(2/3)(1/3)^2) \tag{25}$$

$$b^3 = 6/27(1/3)^3 \tag{26}$$

The frequency of shoot apex types after the second cycle of division and apical cell selection in those apices described by the term $3ab^2$, where $a = 1/3$ and $b = 2/3$, is

$$12/27(a^3 + 3a^2b + 3ab^2 + b^3) \tag{27}$$

$$a^3 = 12/27(1/3)^3 \tag{28}$$

$$3a^2b = 12/27(3(1/3)^2(2/3)) \tag{29}$$

$$3ab^2 = 12/27(3(1/3)(2/3)^2) \quad (30)$$

$$b^3 = 12/27(2/3)^3 \quad (31)$$

Thus, after two cycles of mitosis and apical cell selection, the frequency of apices that have fixed the mutation is (19) + (23) + (28), or 11.9 percent, whereas the frequency of apices that have lost the mutation is (20) + (26) + (31), or 43.6 percent. The frequency of apices that still remain a chimera is 44.5 percent, although it should be remembered that these apices consist of two types, those that have two mutant apical initials and one wild type and those with one mutant apical initial and two wild types. These two types will occur in a 1:2 ratio. Thus, with repeated cycles of division and random apical initial selection, the frequency of apices that have fixed the mutation approaches 1/3 or 1/n, and the frequency of apices that have lost the mutation approaches 2/3 or n–1/n.

It should be noted that the binomial solution presented here for the fixation of mutations in shoot apices with different numbers of apical initials is only an approximate solution to the problem. A more realistic approach involving stochastic processes and Markov chains will be presented by Drs. L. and N. Fukshansky and the author in a future communication. The final probabilities of mutation fixation and loss regardless of the method of analysis are still n and n–1/n.

Extrapolating from these examples, the probability of mutation fixation in a shoot apex is related to the average number of apical initials in the apex. Li and Rédei (1969) pioneered this area, and their work on *Arabidopsis* prompted the calculation in the present paper. These researchers have shown that *Arabidopsis* seeds exposed to mutagens often produce plants that are a chimera; some flowers upon selfing showing a 4:0 ratio and others a 3:1 for an induced mutation. Since the seed contains an embryo with a small shoot apex, they have interpreted these results in terms of a germ line in the seed that consists of two cells. The number of cells in the germ line is called the genetically effective cell number (GECN) (Li and Rédei, 1969).

In seed plants the concept of distinct cells in the seed that serve as the basis of a germ line in the development of meiotic tissue is not the typical view of early seedling development. In individual angiosperm shoot apices, there may be a group of nondividing cells that are carried along and do not divide until reproductive development is initiated (Buvat, 1955; Steeves and Sussex, 1972). These nondividing cells comprise the *méristème d'attente*. The *méristème d'attente* cannot account for a germ line transmitted in a branching system, since the development of axillary buds (the means of branching in seed plants) is from meristematic, not quiescent, cells. The development of the *méristème d'attente* is a phenomenon of the more mature vegetative shoot apex. Thus, although *Arabidopsis* may have a germ line (GENC = 2) in the seed, this is not necessary (as the previous discussion has shown). A hypothesis based upon the random selection of two apical initials, on the average, after each cycle of division will give the same results. The random selection of apical initials also would occur in the development of axillary buds in seed plants, which are large and morphologically more complex than *Arabidopsis.* Because of the association of GENC with a germ line (Rédei, 1975), the present paper will continue to refer to the average number of apical initials (n) in a shoot apex, since this more closely approximates apical shoot anatomy and ontogeny.

Within a shoot apex mutation induction and fixation can be viewed in the following way:

$$\text{Mutation} \times \text{Fixation} = nA\left(\frac{1}{n}\right) = A \tag{32}$$

$$\text{Mutation} \times \text{Loss} = nA\left(1 - \frac{1}{n}\right) = nA - \frac{nA}{n} \tag{33}$$

$$\text{Nonmutation} \times \text{Fixation} = (1 - nA)\left(\frac{1}{n}\right) = \frac{1}{n} - \frac{nA}{n} \tag{34}$$

$$\text{Nonmutation} \times \text{Loss} = (1 - nA)\left(1 - \frac{1}{n}\right)$$

$$= 1 - \frac{1}{n} - nA + \frac{nA}{n} \tag{35}$$

thus the probability of mutation and fixation is

$$\frac{A}{A + nA - \frac{nA}{n} + \frac{1}{n} - \frac{nA}{n} + 1 - \frac{1}{n} - nA + \frac{nA}{n}} = A \tag{36}$$

Therefore, regardless of the number of apical initials, as long as there is no selection between initials the frequency of mutant shoot apices is the mutation probability per apical initial. To calculate the size of the population of shoot apices (N) necessary to find at least one apex that has fixed the mutation with an assurity of 95 percent, one can again use a binomial expansion, $(A+(1-A))^N = 1$,

$$(A+(1-A))^N - (1-A)^N = .95$$

$$N = \frac{\log .05}{\log (1-A)} \tag{37}$$

An important aspect influencing the probability of fixation is selection within the shoot apex. In shoot apices with more than one apical initial, the mutations that are screened should be selectively neutral in terms of competition between apical initials. Thus, only true recessives and/or mutations of genes not expressed within the apical shoot environment should be monitored. Any mutation that lessens the growth rate of the apical initial will drastically decrease the frequency of shoot apices that have fixed the mutation. Whether the same applies to shoot apices with a single apical cell is unknown. The question of whether an apical cell is replaced when its biochemistry is quantitatively impaired is unresolved. The physical destruction of an apical-cell-based shoot apex does result in the regeneration of new apices from the uninjured tissue (Steeves and Sussex, 1972; Wardlaw, 1965). Presumably mutations that seriously impair the growth of an apical cell will result in the regeneration of a new apical cell from surrounding cells but one might suspect that in general apical-cell-based meristems would be less responsive to cell selection.

Thus to maximize the sensitivity of plant in situ bioassays, species possessing a single apical initial per shoot apex should be monitored to lessen the loss of mutations through interinitial competition. Also, broad criteria of genetic damage should be selected to increase the value of A. The latter is important because if mutant phenotypes can be caused by a number of different kinds of mutations, the overall frequency of such phenotypes will be the sum of the probabilities of the individual mutations.

PLANT MORPHOLOGY AND MUTAGENESIS

The previous discussion dealt with the problem of detecting shoot apices that have fixed somatic mutations. Having established the frequency of shoot apices that segregate for the mutation in a sample of shoot apices, this information is now used to calculate the incidence of mutation. This calculation is based upon an understanding of the pattern and arrangement of shoot apices on the vascular plant species studied. The significant point is to discriminate between independent mutations versus the clonal spread of a single mutation. For example, if 10 mutant shoot apices are detected in a sample of 100 shoot apices, does this represent 10 different mutations or the vegetative spread of a single mutation? Superficially this appears easy, only a matter of determining whether the shoot apices originated from one individual or 10. In natural populations of plants, the discrimination of clones and individuals is often difficult.

Populations of plants may be of two kinds: sexual populations, in which each individual originates from a distinct zygote, and clonal populations, where large numbers of individuals are descended from a single zygote. In sexual populations where one shoot apex is sampled per plant, the calculation of mutation frequency is straightforward. In clonal populations such calculations are more complex.

Clonal populations can originate through a diversity of means, including various forms of vegetative reproduction (Bell and Tomlinson, 1980; Stebbins, 1950) and various apomictic alterations of the sexual cycle (Swanson, 1957, for review). Regardless of the clonal process, a shoot apex that has fixed a mutation may give rise to a family of plants in which each carries the mutation. The dynamics of mutation accumulation during clone development can best be understood by first considering a simple dichotomous rhizome such as found in the fern *Osmunda regalis*. In such a rhizome system, if n is the number of apical dichotomies, then the number of shoot apices after each dichotomy follows the pattern,

$$2^1 = 2$$

$$2^2 = 4$$

$$2^3 = 8$$

.

.

.

$2^n = Y_T$, the total number of shoot apices. If R equals mutations fixed per new shoot apex, then after the first dichotomy, n = 1,

the number of mutant apices = $2^n R$

and the number of nonmutant apices = $2^n - 2^n R = 2^n(1-R)$

after the second dichotomy, $n = 2$,

$$\text{the number of mutant apices} = 2^nR + 2^n(1-R)R$$

$$\text{and the number of nonmutant apices} = 2^n(1-R) - 2^n(1-R)R = 2^n(1-R)(1-R)$$

and after the third dichotomy, $n = 3$,

$$\text{the number of mutant apices} = 2^nR + 2^n(1-R)R + 2^n(1-R)(1-R)R$$

$$\text{and the number of nonmutant apices} = 2^n(1-R)^3$$

thus the number of nonmutant apices after n apical dichotomies is

$$Y_{nm} = 2^n(1-R)^n \quad (38)$$

To calculate the frequency of nonmutant shoot apices,

$$F_{nm} = \frac{\text{Number of nonmutant shoot apices}}{\text{Total number of shoot apices}}$$

$$F_{nm} = \frac{2^n(1-R)^n}{2^n} = (1-R)^n \quad (39)$$

These equations can be easily amended to reflect different starting points. For example, if one wants to consider the impact of mutation after a clone has N apices, then,

$$Y_{nm} = N2^n(1-R)^n \quad (40)$$

$$F_{nm} = \frac{N2^n(1-R)^n}{N2^n} = (1-R)^n \quad (41)$$

or if a population of clones, P, each consisting of N apices, then the impact of mutation from a given point in time is

$$Y_{nm} = PN2^n(1-R)^n \quad (42)$$

$$F_{nm} = \frac{PN2^n(1-R)^n}{PN2^n} = (1-R)^n \quad (43)$$

These ideas of clone growth and mutation spread can be extended to generate equations to describe the entire spectrum of clone types in plants. These equations are based upon the concept of doubling time, D, the time required for the number of shoot apices in a population (clone) to double. Doubling time allows the application of equations used in describing exponential growth in organisms such as bacteria.

Since the total number of apices in a population of clones is

$$Y_T = C2^n \tag{44}$$

where C is a constant and equal to NP and n is the number of cycles of dichotomous divisions. This growth is also described by

$$Y_T = Ce^{kt} \tag{45}$$

where t equals time (in years), e is the base of natural logarithms, and

$$k = \frac{\ln 2}{D} \tag{46}$$

Thus, for a given clone, since the number of nonmutant apices is

$$Y_{nm} = 2^n(1-R)^n \tag{47}$$

then

$$Y_{nm} = e^{kt}(1-R)^n \tag{48}$$

since

$$n = \frac{kt}{\ln 2} \text{ and } e = 2^{1/\ln 2} \tag{49}$$

then

$$Y_{nm} = 2^{kt/\ln 2}\,(1-R)^{kt/\ln 2} \tag{50}$$

The frequency of nonmutant apices is

$$F_{nm} = \frac{Y_{nm}}{e^{kt}} = (1-R)^{kt/\ln 2} \tag{51}$$

The frequency of mutant apices is

$$F_m = 1 - F_{nm} \tag{52}$$

Table 4.2 gives the F_m values per clone for different forward mutation values, R, different doubling times, D, for different aged clones. Three values of forward mutation, R, are given. These values are not for individual genes but rather for a group of genes that contribute to measurable genetic damage in terms of readily scored mutant phenotypes. The results in Table 4.2 suggest two important strategies in the development of sensitive in situ mutagen bioassays. Again, mutant phenotypes that can result from any one of a number of gene mutations should be selected to increase the value of R, and plants with fast growth rates in terms of multiplication of shoot apices to increase the value of D.

In sampling a population that is a mosaic of clones, the mutation frequency per clone can be calculated by using the zero term of a Poisson distribution. A uniform number of apices from each clone is screened for the mutation segregation pattern. Clones are scored as to whether they have one or more mutant shoot apices or are mutant

TABLE 4.2
Frequency of Mutant Shoot Apices in Clones of Different Ages (t) with Different Mutation Rates (R) and Doubling Times (D)

		Doubling Time (D)			
	t	*0.25*	*0.5*	*1.0*	*10*
$R = 10^{-3}$	1	.004	.002	.001	.000
	10	.039	.020	.007	.001
	50	.181	.096	.034	.005
	100	.330	.181	.068	.010
	500	.865	.632	.300	.049
$R = 10^{-4}$	1	.000	.000	.000	.000
	10	.004	.002	.001	.000
	50	.020	.010	.005	.001
	100	.039	.020	.010	.001
	500	.181	.095	.049	.005
$R = 10^{-5}$	1	.000	.000	.000	.000
	10	.000	.000	.000	.000
	50	.002	.001	.001	.000
	100	.004	.002	.001	.000
	500	.020	.010	.005	.001

Note: R equals mutations fixed per shoot apex, t is the age in years, and D is the time in years necessary to double the population of shoot apices.

free. The clones that have more than one mutant shoot apex may represent either a single mutation or multiple mutations. It is assumed that the mutations are distributed independently at different shoot apices, so that the frequency of clones with 0, 1, 2, 3 · · · mutant shoot apices will be given by successive terms in a Poisson distribution with a mean m, the average number of different mutations per clone. The frequency of mutant-free clones (F) is calculated

$$F = e^{-m} \tag{53}$$

$$m = -\ln F \tag{54}$$

The value m is calculated for each population of clones studied. The mean frequency of mutations per clone (m) is actually the mutation frequency per individual plant (of sexual origin); thus, different plant species, as well as populations of the same species growing in different environments, can be compared.

Another aspect of clonal reproduction and mutagenesis is the possibility of dating mutations. For plants with regular patterns of branching and known growth rates, the distribution of somatic mutations among related shoot apices can be used to construct a phylogeny of the mutations. Analysis of the distribution of chromosome mutations among the shoot apices of the royal fern (*Osmunda regalis*) allowed preliminary attempts at dating the mutations (Klekowski and Berger, 1976). The results of these studies indicated that 64 percent of the mutations occurred five years prior to the time of the study. The potential of in situ mutagen assays in detecting and dating episodes of mutagenic pollution has not been researched adequately.

FERNS: IN SITU MUTAGEN BIOASSAYS

Pteridophytes have morphological and anatomical features that enhance their sensitivity as bioassays in contrast with seed plants. The life cycle of most ferns consists of two distinct free-living forms, the diploid sporophyte and haploid gametophyte. The sporophyte is the portion of the life cycle that is used to screen an environment. Sporophyte organs include rhizomes (stems), sterile and fertile fronds (leaves), and roots. The rhizomes terminate in shoot apices that have apical meristems based upon single apical cells. Thus mutations fixed in these shoot apices are not lost as readily through competition between several apical initials in one apex.

The fern genotype appears to have genes that are expressed in the sporophyte, in the gametophyte, or in both generations. The compartmental nature of the fern genotype can be advantageously manipulated to reveal a broad spectrum of forward mutations. Figure 4.1 depicts this life cycle in terms of where mutations can be detected. Three portions of the life cycle have the greatest utility in this regard: meiosis, early gametophyte development, and zygote formation. In many ferns meiosis is particularly suitable for screening two break chromosome mutations as well as micronuclei (see Ma, 1979, for a discussion of meiotic micronuclei in plants).

The gametophytes of many ferns are readily cultured on defined medium, so screening for aberrant phenotypes is relatively easy. A common mutant phenotype consists of gametophytes that are capable of only two or three mitotic divisions. The mutant is composed of a few prothallial cells and a rhizoid. This mutant phenotype has terminated growth, but will remain alive for at least three more weeks. The phenotype may be the result of a mutation in a gene that is expressed in both the gametophyte and the sporophyte generations. Physiological studies in ferns support the idea that early germination activity of fern spores is based upon preformed proteins and RNA templates from genes expressed in the premeiotic spore mother cell (DeMaggio and Raghavan, 1973; Raghavan, 1980). The pattern of development of this mutant phenotype conforms to the idea that the spore mother cell is heterozygous for both wild-type and mutant alleles, and at least the wild-type allele is transcribed; the gene products are partitioned randomly into the cytoplasm of the quartet of spores at meiosis. Thus, early germination events are

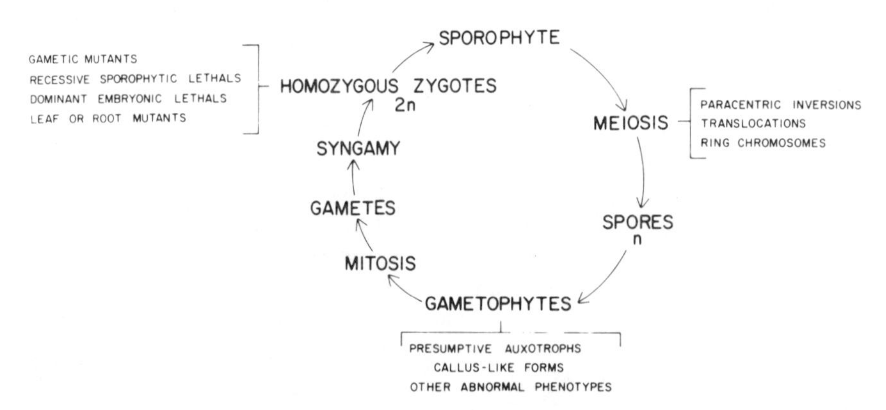

FIGURE 4.1. Manipulation of the Homosporous Fern Life Cycle to Reveal Various Categories of Mutations.

normal in all spores, but when gametophyte development becomes dependent upon genes transcribed in the gametophyte nuclei, the mutants terminate their growth. This gametophyte phenotype represents a forward mutation in genes that are critical for early spore germination and are transcribed premeiotically in the spore mother cell as well as in the gametophyte.

Another common gametophyte phenotype is a callus-like form that has lost the capacity for organized two-dimensional planar development. Such phenotypes have also been induced with ionizing radiation (Partanen, 1958). Gametophyte mutations that result in altered chloroplast distribution in the cells have been induced with ionizing radiation in *Osmunda regalis* (Haigh and Howard, 1970), and a sporophyte segregating for a similar phenotype has been discovered in *Acrostichum* (Lloyd, personal communication). In addition to mutations resulting in aberrant gametophyte morphologies, it is possible also to screen for gamete mutations (Duckett, Klekowski, and Hickok, 1979).

The final category of detectable mutations is the one that influences sporophyte development (Figure 4.1). The homosporous fern life cycle is characterized by complete and obligate homozygosity after one generation of selfing (intragametophytic selfing) (Klekowski and Lloyd, 1968). Isolated gametophytes often form both eggs and spermatozoids; since gametes in vascular plants arise through mitotic cell divisions, the gametes produced by a single gametophyte have identical genotypes. Thus, selfing a gametophyte screens the entire genotype for recessive deleterious mutations that are expressed in the zygote and young embryo. Such sporophytic genetic load estimates have been conducted in a number of species. See Klekowski (1970) for the diversity of aberrant sporophyte phenotypes that have been detected in genetic load studies and Klekowski (1973; 1976a) for limited genetic analysis of the mutants.

Because of the above characteristics (apical meristems with single apical cells and ease of measurement of forward mutations) ferns have been successfully used as in situ mutagen bioassays (Klekowski, 1978). A diversity of mutational criteria has been tested in these studies, including chromosome mutations (Klekowski and Berger, 1976; Klekowski and Levin, 1979), sporophytic mutations (Klekowski, 1976b; Klekowski and Davis, 1977), and gametophytic mutations (Klekowski and Klekowski, 1982). The class of mutations that shows the greatest promise for future bioassay use is the gametophytic mutations. These mutations can be screened for relatively quickly and cheaply, with the added bonus that their genetic basis is readily ascertained. Spore collections from individual sporophytes are sown on agar-solidified nutrient medium, then grown under continuous fluorescent illumination (400 foot-candles); gametophyte cultures are screened for morphologically aberrant forms. If the parental sporophyte is heterozygous for the mutant, 1:1 ratios usually are encountered in the culture. This technique works only with fern species that lack antheridogens. Antheridogen systems (Näf, 1979) usually will mask the presence of mutant gametophyte phenotypes.

An example of the use of ferns as in situ mutagen bioassays based upon measurements of gametophyte mutations in clones has been completed by Klekowski and Klekowski (1982). The ostrich fern (*Matteuccia struthiopteris*), a common component of northeastern American riparian environments, was the bioassay organism. This fern forms clonal populations of erect, frond-bearing rhizomes interconnected by underground stolons (Lloyd, 1971). The erect rhizomes form photosynthetic fronds as well as fertile fronds that mature in the winter. The apical meristems of the erect rhizomes have a single apical cell (Wardlaw, 1965).

TABLE 4.3
Somatic Mutations in *Matteuccia struthiopteris* Populations in Massachusetts

				Mutations Detected		
Population	*Clones*	*Erect Rhizomes*	*Mutant-free Clones*	*1:1*	*Other**	*Percent*
Connecticut River	19	234	15	4	—	1.71
Deerfield River	15	193	13	2	—	1.44
Green River	12	93	9	2	1	3.22
Mill River, Northampton	6	100	4	2	2	4.00
Oxbow, Northampton	2	33	2	—	—	0.00
Total Controls	54	653	43	10	3	1.99
Housatonic River	23	199	6	5	15	10.05

*Presumed chromosome mutations. (See text for discussion.)

Source: Klekowski and Klekowski, 1982.

The mutation frequency per clone was determined for populations growing either in an environment contaminated with polychlorinated biphenyls (PCBs) or in uncontaminated environments. Spore samples from erect rhizomes of different clones were screened for gametophyte mutations. Clones were classed as either mutant or mutant-free. Table 4.3 shows these data. The PCBs-contaminated populations are growing in the flood plain of the Housatonic River downstream from the city of Pittsfield, Massachusetts. The control populations are growing in the Connecticut River valley or its tributaries in Massachusetts. The samples had two general classes of mutations. Spore samples generated gametophyte cultures that had two distinct kinds of morphs in a 1:1 ratio or that had mutant gametophytes with various aberrant morphologies. The latter usually did not occur in 1:1 ratios; depending upon the spore sample, 15 percent to 70 percent of the gametophytes were aberrant, and were hypothesized to result from erect rhizomes heterozygous for chromosome aberrations (inversions, translocations), although this has not yet been confirmed with cytological investigation. The Housatonic had a high frequency of such spore samples.

The frequency of mutations per clone was calculated, using the zero term of the Poisson distribution (see [53]). For the Housatonic population the number of clones studied was 23, and the number without mutations was 6 (Table 4.3). Thus

$$e^{-m} = .2609 \tag{55}$$

$$m = 1.3436 \tag{56}$$

For the five control populations the total number of clone samples was 54, of which 43 were mutant-free (Table 4.3). Thus

$$e^{-m} = .7963 \tag{57}$$

$$m = .229 \tag{58}$$

Therefore the mean frequency of mutations per clone (m) in the total controls was .229, whereas in the Housatonic it was 1.3436, a 5.87-fold increase.

The mutation frequency also can be calculated by relating the number of different mutations to the total number of apices (erect rhizomes) sampled. This procedure underestimates the mutation frequency, since it often is difficult to determine the number of different mutations that occurred in a clone. This is especially a problem where the mutant phenotype can arise from a diversity of causes (different genes mutating). Applying these procedures to the data in Table 4.3, the total number of erect rhizomes in the controls was 653, with 13 different mutations occurring (1.99 percent). In the Housatonic, 199 erect rhizomes were sampled, with 20 different mutations found (10.05 percent). Thus the frequency of mutations in the Housatonic was 5.05 times the controls. This value is remarkably similar to the 5.87-fold difference calculated using equation (53). It should be noted that in the former case the mutation frequency was based upon the frequency of erect rhizomes sampled, whereas the latter was based upon the number of clones. Because of difficulties in distinguishing different mutants, 5.87 is probably closer to the actual value.

The studies on ferns as in situ mutagen assays show that the genetic impact of ambient concentrations of mutagenic and/or carcinogenic pollutants can be detected. The significance of these studies is that they illustrate the need to incorporate basic botanical knowledge in the development of plant-based mutagen bioassays. Further researches in the development of plant-based in situ mutagen bioassays must consider the anatomy, morphology, and ontogeny as well as the genetics of the bioassay species.

ACKNOWLEDGMENTS

This paper was written during my tenure as an Alexander von Humboldt Stipendiat at the Biological Institute, University of Freiburg, Freiburg, West Germany. I should like to express my appreciation to Professor Hans Mohr for his gracious hospitality.

REFERENCES

Bell, A. D., and P. B. Tomlinson. 1980. Adaptive architecture in rhizomatous plants. *Bot. J. Linn. Soc.* 80:125-160.

Bierhorst, D. W. 1977. On the stem apex, leaf initiation and early leaf ontogeny in filicalean ferns. *Amer. J. Bot.* 64:125-152.

Buvat, R. 1955. Le méristème apical de la tige. *Ann. Biol.* 31:595-656.

DeMaggio, A. E., and V. Raghavan. 1973. Photomorphogenesis and nucleic acid metabolism in fern gametophytes. In *Advances in Morphogenesis*, ed. M. Abercrombie, J. Brachet, and T. J. King. Vol. 10. New York and London: Academic Press.

Duckett, J. G., E. J. Klekowski, and L. G. Hickok. 1979. Ultrastructural studies of mutant spermatozoids in ferns. I. The mature nonmotile spermatozoid of mutation 230X in *Ceratopteris thalictroides* (L.) Brongn. *Gamete Research* 2:317-343.

Gifford, E. M., V. S. Polito, and S. Nitayangkura. 1979. The apical cell in shoots and roots of certain ferns: A re-evaluation of its functional role in histogenesis. *Plant Sci. Letters* 15:305-311.

Haigh, M. V., and A. Howard. 1970. Mutations affecting cell morphology in *Osmunda regalis. J. Hered.* 61:285-287.

Johnston, J. B., and P. K. Hopke. 1980. Estimation of the weight-dependent probability of detecting a mutagen with the Ames assay. *Environ. Mutagenesis* 2:419-424.

Klekowski, E. J., Jr. 1970. Populational and genetic studies of a homosporous fern—*Osmunda regalis. Amer. J. Bot.* 57:1122-1138.

——. 1973. Genetic load in *Osmunda regalis* populations. *Amer. J. Bot.* 60: 146-154.

——. 1976a. Genetics of recessive lethality in the fern *Osmunda regalis. J. Hered.* 67:146-148.

——. 1976b. Mutational load in a fern population growing in a polluted environment. *Amer. J. Bot.* 63:1024-1030.

——. 1978. Detection of mutational damage in fern populations: An *in situ* bioassay for mutagens in aquatic ecosystems. In *Chemical Mutagens. Principles and Methods for Their Detection,* ed. A. Hollaender and F. de Serres, vol. 5, pp. 79-99. New York: Plenum Press.

Klekowski, E. J., Jr., and B. B. Berger. 1976. Chromosome mutations in a fern population growing in a polluted environment: A bioassay for mutagens in aquatic environments. *Amer. J. Bot.* 63:239-246.

Klekowski, E. J., Jr., and E. L. Davis. 1977. Genetic damage to a fern population growing in a polluted environment: Segregation and description of gametophyte mutants. *Can. J. Bot.* 55:542-548.

Klekowski, E. J., Jr., and E. Klekowski. 1982. Mutation in ferns growing in an environment contaminated with polychlorinated biphenyls. *Amer. J. Bot.* (in press).

Klekowski, E. J., Jr., and D. E. Levin. 1979. Mutagens in a river heavily polluted with paper recycling wastes: Results of field and laboratory mutagen assays. *Environ. Mutagenesis* 1:209-219.

Klekowski, E. J., Jr., and R. M. Lloyd. 1968. Reproductive biology of the Pteridophyta. I. General considerations and a study of *Onoclea sensibilis* L. *Bot. J. Linn. Soc.* 60:315-324.

Li, S. L., and G. P. Rédei. 1969. Estimation of mutation rate in autogamous diploids. *Radiation Botany* 9:125-131.

Lloyd, R. M. 1971. Systematics of the onocleoid ferns. *Univ. Calif. Publ. Bot.* 61:1-93.

Lower, W. R., P. S. Rose, and V. K. Drobney. 1978. *In situ* mutagenic and other effects associated with lead smelting. *Mutation Res.* 54: 83-93.

Ma, T. H. 1979. Micronuclei induced by X-rays and chemical mutagen in meiotic pollen mother cells of *Tradescantia. Mutation Res.* 64:307-313.

Näf, U. 1979. Antheridogens and antheridial development. In *The Experimental*

Biology of Ferns, ed. A. F. Dyer, pp. 254-306. London: Academic Press.

Partanen, C. R. 1958. Quantitative technique for analysis of radiation-induced tumorization in fern prothalli. *Science* 128:1006-1007.

Raghavan, V. 1980. Cytology, physiology, and biochemistry of germination of fern spores. *Int. Rev. Cytology* 62:69-118.

Rédei, G. P. 1975. Induction of auxotrophic mutations in plants. In *Genetic Manipulations with Plant Material*, ed. L. Ledoux, pp. 329-350. New York: Plenum Press.

Stebbins, G. L. 1950. *Variation and Evolution in Plants.* New York: Columbia University Press.

Steeves, T. A., and I. M. Sussex. 1972. *Patterns in Plant Development.* Englewood Cliffs, N.J.: Prentice-Hall.

Stewart, R. N. 1978. Ontogeny of the primary body in chimeral forms of higher plants. In *The Clonal Basis of Development*, ed. S. Subtelny and I. M. Sussex, pp. 131-160. New York: Academic Press.

Swanson, C. P. 1957. *Cytology and Cytogenetics.* Englewood Cliffs, N.J.: Prentice-Hall.

Wardlaw, C. W. 1965. *Organization and Evolution in Plants.* London: Longmans, Green.

5

The Use of Macrophytic Algae in Bioassays for Mutagens Dispersed in the Marine Environment

William S. Barnes

CARCINOGENS IN THE MARINE ENVIRONMENT

The presence of pollutants in the sea has been a source of concern for some time; however, the probability that many of these pollutants are chemical carcinogens and mutagens has only recently become an issue. The reasons for such concern are basically twofold. First and most important is the impact that these carcinogens might have on the human population through exposure via contaminated seafood. For example, mussels and other shellfish are known to concentrate benzo(a)pyrene and other potentially carcinogenic polycyclic aromatic hydrocarbons (PAH) from seawater (Gesamp, 1977). Most species of fish seem capable of metabolizing PAH but do accumulate low-molecular-weight aromatic hydrocarbons (naphthalene and its metabolites) and polychlorinated biphenyls (Malins, 1980). Another concern, less frequently expressed, is that damage to the genetic systems of various elements of the biota might contribute to a reduction of vitality or reproductive ability.

Carcinogens in Rivers

Partial evidence for the presence of carcinogens in the marine environment comes from chemical analysis of river systems that are an important source of carcinogen and mutagen input. Neff (1979) has estimated that approximately 21 percent of benzo(a)pyrene (B(a)P) in the oceans is introduced from domestic and industrial wastes and surface runoff from land. Most of the remainder comes from fallout and rainout from air, with only a minor component (4 percent) due to spillage and transport operations. It might be expected that river systems are a major route of entry for other chemical carcinogens and mutagens as well. Some of the organic chemicals found in heavily polluted U.S. rivers are shown in Table 5.1, and include proven and suspected carcinogens. These data imply that a variety of carcinogens are entering marine environments. An extensive discussion of this situation can be found in Neff (1979), Kraybill (1976), and Cole (1979).

TABLE 5.1
Organic Pollutants Found in U.S. Rivers

Fox River, Chicago, Illinois (from Brown et al., 1973)		Kanawha River, West Virginia (from Kraybill, 1976)	
ethyl ether		carbon tetrachloride	+
crude oil		ethylenethiourea	?
gasoline		vinyl chloride	+
benzanthracene	+	polyurethane	
organophosphates		chloroform	+
toluene		benzene	?
benzene	?	bis(2-chloroethyl)ether	?
naphthalene			
benzoic acids			
triazines	?*		
toluidine			
chlorinated hydrocarbons	+*		
Charles River, Boston, Massachusetts (from Hites and Biemann, 1972)		**Buffalo River, Buffalo, New York (from Nelson and Hites, 1980)**	
alkanes		1-naphthylamine	?
naphthalenes		nitrobenzene	
anthracenes		2,4-dinitrotoluene	
phenanthrenes		2,6-dinitrotoluene	
pyrene		trichlorobenzenes	
fluoranthene			
dibutyl phthalate			
di(2-ethylhexyl)phthalate			

+ Proven carcinogen in humans or animals.

? Suspect carcinogen in humans or animals.

* Only some of these compounds have the activity indicated.

Note: Proven carcinogens are those compounds for which there is "sufficient" evidence of human carcinogenicity or which were positive in at least two studies as determined by IARC expert committees. Suspect carcinogens are positive in at least one animal study or are on the EPA list of suspected carcinogens (EPA, 1976).

Pollution-Related Neoplasms

Another line of evidence suggesting the presence of carcinogens in marine environments is the appearance of tumors in fish and bivalve mollusks. This may indicate that carcinogenic organics are present in biologically significant concentrations. It has long been suspected that the distribution of such tumors may be correlated with the distribution of pollution, but in fish it has been difficult to fix a cause for unusually high incidences of tumors. The best example is probably a study by Brown et al. (1973) in which 16 species of fish were compared in clean and highly polluted environments. Fish in the polluted Fox River, Chicago, watershed exhibited a higher incidence of tumors than the same species in a clean Canadian watershed. Other studies have linked tumors in hagfish from a Swedish fjord and northern pike from the Baltic Sea with an increase or decrease in pollutant levels (Falkner et al., 1976; Ljunberg, 1976).

There is an interesting correlation between invasive neoplasias in soft-shell clams (*Mya arenaria*) and a spill of JP5 jet fuel and no. 2 fuel oil at Searsport, Maine (Barry and Yevich, 1975). An instance has also been reported of "cancerous disease" in *Porphyra tenera* in water receiving effluents from a coal chemical plant at Ohmuta, Japan (Ishio, Kawabe, and Tomiyama, 1972a; Ishio, Nakagawa, and Tomiyama, 1972b), although this may be attributable to a teratogenic rather than a carcinogenic effect (Gesamp, 1977). For a further discussion of pollution-related neoplasms, see Dawe, Scarpelli, and Wellings (1976); Stich and Acton (1976); and Kraybill et al. (1977).

Carcinogens in Bivalves

A number of studies on mussels (*Mytilus* spp.) from areas subject to oil pollution have shown that these organisms are capable of concentrating hydrocarbons from seawater. Among these hydrocarbons are the known carcinogens benz(a)anthracene and benzo(a)pyrene. In addition, a large number of aromatic hydrocarbons are known or suspected carcinogens. Table 5.2 summarizes some of the data available on the accumulation of hydrocarbons by *Mytilus* spp. For reviews, see Neff (1979) and Lee (1977).

There seems to be sufficient evidence to indicate the existence of potentially hazardous carcinogenic contamination in the sea. If this is true, it would be prudent to identify contaminated areas in order to minimize human exposure in the short term and to eliminate the situation in the long term. This problem is only beginning to receive attention, and consequently there are no generally accepted methods for the detection of mutagens and carcinogens in marine environments. Indeed, there are hardly any data with which the practical utility of any technique can be assessed. The state of the art with regard to the use of macrophytic algae is no further advanced than any of the other approaches, and therefore much in this chapter must necessarily be somewhat conjectural. Bearing in mind this limitation, however, I will discuss the advantages and

TABLE 5.2
Concentrations of Petroleum Hydrocarbons in *Mytilus* spp. from Contaminated Environments

Contamination Source	*Hydrocarbons Analyzed*	*Hydrocarbon Concentration in Animal (μg/g Wet Tissue)*	*Reference*
Spill—bunker C	aromatics	77-103	Lee (1977)
Spill—fuel oil no. 2	n-paraffins	1.4	Lee (1977)
Chronic	aromatics	6-75	Lee (1977)
Chronic	n-paraffins	1.0	Lee (1977)
Spill	aliphatics and aromatics	400	Lee (1977)
Chronic	aromatics	8	Lee (1977)
Chronic	aliphatics	250	Lee (1977)
Chronic	benzo(a)pyrene	0-42	Dunn and Stich (1975)
Chronic	aliphatics	8-220	Fossato and Siviero (1974)
Spill—fuel oil no. 2	benz(a)anthracene benzo(a)pyrene	0.2-0.6	Pancirov and Brown (1977)
Chronic	benzo(a)pyrene	12-135	Fossato, Nasci, and Dolci (1979)

disadvantages of using macrophytic algae in bioassays for marine mutagens and carcinogens, and some of the techniques that can be used to exploit their attributes.

DETECTION OF POTENTIALLY CARCINOGENIC MARINE POLLUTANTS USING ALGAL MACROPHYTES AND AN IN VITRO ASSAY FOR GENETIC ACTIVITY

Discussion of the Technique

Potential carcinogens dispersed in marine environments will probably be present in rather small concentrations. It follows that a concentration step will be needed for the most efficient detection of genetic activity. One way in which this can be accomplished is through the use of bioconcentrators. Extracts of such bioconcentrators, enriched in a variety of contaminants, can then be assayed at different genetic end points, using appropriate short-term assays. A similar strategy has often been employed with mussels and other shellfish, but with the difference that a chemical analysis is performed. This type of assay has reached a sufficient degree of sophistication that a world-wide monitoring program (Musselwatch) is now following the long-term concentrations of various heavy metal and chemical pollutants in mussels.

The attraction of this approach is that organisms that may be good bioconcentrators, but are intractable for genetic analysis, can be coupled with established laboratory assays for mutation, gene conversion, mitotic crossing-over, nondisjunction, and chromosome aberration (see "Testing for Genetic Activity," below).

There are several other advantages that recommend this approach rather than the alternative of chemical concentration or chemical analysis.

1. The occurrence of pollutant events may be episodic in both time and place. The use of chemical or physical concentration methods necessitates "point" sampling at a single time and place, and this procedure may give misleading estimates of long-term trends. Of course, it is possible to circumvent this problem by using a more intricate sampling procedure, but the complexity of the experimental design is then correspondingly increased. Biological concentrators, on the other hand, give a moving time-averaged estimate of pollutant levels, and sampling schemes can therefore be greatly simplified.
2. The expense for supplies, equipment, and sampling logistics will be greater with chemical and physical methods. At the present time lyophilization (Simmon and Tardiff, 1976) or adsorption onto nonpolar resins (Glatz et al., 1978; Cheh et al., 1980; Nestmann et al., 1979) has been most frequently used. In comparison with the initial outlay and operating expense for pumps, resins, collection apparatus, or freeze-dryers, the collection of seaweed is quite inexpensive.
3. The detection of mutagenic substances in or on macrophytic algae gives a direct indication of biological significance, since it shows that the substance may be biologically concentrated or at least accumulated, and it suggests a route of entry into the food chain. Furthermore, if the substance is found inside the cells, the capacity to cross cytoplasmic, and presumably nuclear, membranes is demonstrated. The potential for such chemical-biological interactions cannot be inferred from studies utilizing only chemical or physical methods.

Disadvantages of this approach include the following:

1. It may not be possible to find suitable bioconcentrators at every site from which a sample is required (see "Properties of Ideal Bioconcentrators," below).

2. Biological concentration is not as "clean" as chemical concentration, and the extract will consist of a complex mixture of biological materials and pollutant. This fact will make chemical analysis and identification of the genetically active compounds more difficult. This is one reason that the use of bioconcentrators is seen primarily as a prescreen for areas of potential hazard, and as a method for identifying the chemical characteristics of potential carcinogens. This information can then be used to direct more intensive sampling and chemical analysis.
3. There are certain attributes that useful bioconcentrators must possess (see following section). More research is still needed to determine how well macrophytic algae may conform to these requirements.

Properties of Ideal Bioconcentrators

General Attributes

Phillips (1977) has listed a number of desirable attributes for organisms used in bioassays:

1. The organism should be abundant, so that collection is facilitated and specimens can be found at every sampling station.
2. The organism should be tolerant of high pollution levels.
3. The organism should be sedentary or sessile, so that it is representative of the area from which it is collected.
4. The organism should be of reasonable size, providing adequate tissue for analysis.
5. The organism should tolerate brackish water. This is especially important where inputs from rivers and surface runoff from the land are of concern.
6. The organism should exhibit a high concentration factor and there should be a simple, reproducible correlation between pollutant content in the organism and average pollutant concentration in the immediate surroundings.

Macrophytic algae have been used primarily as bioconcentrators of heavy metals. *Fucus vesiculosus* is the most common bioindicator (Bryan and Hummerstone, 1973; Preston et al., 1972; Phillips, 1979), but *Ascophyllum nodosum* (Haug, Melsom, and Omang, 1974), *Porphyra* spp. (Preston et al., 1972), *Enteromorpha linza* (Seeliger and Edwards, 1977), and *Ulva* spp. (Seeliger and Edwards, 1977) have also been used. All of these algae meet the requirements of abundance and size. *F. vesiculosus*, *Ulva* spp., and *Enteromorpha* spp. occur frequently in brackish water. Relatively few data are available on pollution sensitivity of various algae (Hellenbrand, 1978). Hirose (1978) studied the change in species composition in relation to "pollution" in the Seto Inland Sea. He found that *U. pertusa* and several species of *Enteromorpha* were strongly tolerant. whereas *Ectocarpus siliculosus* and five species of *Sargassum* were classified as weakly tolerant. All species of *Rhodophyceae* were pollution-sensitive. Borowitzka (1972) studied species composition at sewage outfalls and reached similar conclusions; *Ulva* spp. and *Enteromorpha* spp. were especially tolerant. Steele (1978) investigated the effect of oil pollution on *F. edentatus* and *Laminaria saccharina.* Mature *Fucus* plants and *Laminaria* gametophytes were found to be tolerant of no. 2 fuel oil at concentrations in the water of 1–3 ppm. The reproductive stages of both algae were found to be extremely sensitive to inhibition by hydrocarbons, however. The general picture that emerges is that, a priori, *Ulva* and *Enteromorpha* are most suitable for monitoring purposes, while *Fucus, Ascophyllum,* and *Laminaria* may be useful. Concentration of heavy metals and organics in these algae will be discussed later.

Specific Attributes for a Bioconcentrator of Genetic Activity

In addition to the properties mentioned above, there are additional requirements for a bioconcentrator to be used in assays for genetic activity.

1. The organism must not detoxify and/or excrete the mutagenic substance. It is this property that has made the mussel *M. edulis* valuable in monitoring for polycyclic aromatic hydrocarbons (Gesamp, 1977; Malins, 1977). Metabolism of xenobiotics has not been studied in algae. However, it is known that the endoplasmic reticulum, in which the enzymes responsible for the activation and detoxification of mutagens are found, is scanty, if not absent, in the terminally differentiated leaves of higher plants (Heslop-Harrison, 1978). Efforts to find active S-9 fractions in such leaves have, therefore, not been successful. On the other hand, extracts of *Zea mays* seedlings are mutagenic after growth of the seedlings in soil treated with several S-triazine herbicides (Plewa, 1978). Furthermore, there is evidence that soybean seedlings germinated in the presence of dimethylnitrosamine can metabolize this compound to a genetically active species (Arenaz and Vig, 1978). This seems to indicate that very young plants have some ability to activate promutagens, whereas mature plants may not; but whether an analogy can be made with algae is difficult to say.

2. The organism must not produce mutagenic metabolites that would mask or interfere with the detection of mutagenicity due to

contaminants. We have found an example of such interference in the red alga *Plocamium coccineum.* Chloroform-methanol extracts of this alga are directly mutagenic for *Salmonella typhimurium* strain TA100 in the widely used Ames test (Table 5.3). The spectrum of genetic damage is the same as that observed for five acyclic halogenated hydrocarbons that were isolated as natural products from *Plocamium* spp. by Leary et al. (1979). Extracts of *Enteromorpha, Porphyra,* and *Cystoclonium* do not cause an increase in the number of revertants.

3. The organism must not produce toxins that would interfere with the use of in vitro tests.

4. The tests for mutation generally depend on reversion to wild type of a well-defined mutation. Many of these mutations are auxotrophies of some type, and mutants are selected in media lacking the required supplement. It is highly desirable that bioindicators should not contain large pools of these supplements. Contaminating nutrilites can often be removed by partitioning the mutagens into an organic phase. The aqueous phase carrying the nutrilites is discarded (see discussion of techniques, below). In our work we have not found this to be a problem with marine algae.

Concentration of Mutagens and Carcinogens by Marine Algae

Carcinogens and mutagens can be either inorganic or organic compounds. Some of the heavy metals, in certain oxidation states, exhibit mutagenic and carcinogenic activity, and the concentration of heavy metals from the sea by macroalgae has been comparatively well studied. On the other hand, there are very few data available on the ability of these algae to accumulate or concentrate organic compounds.

Heavy Metals

A number of heavy metals are recognized or suspect carcinogens. Arsenic, chromium, cadmium, beryllium, and nickel are considered carcinogenic in humans; beryllium, cadmium, cobalt, chromium, lead, and nickel induce sarcomas and carcinomas in rodents. This subject has been reviewed by Sunderman (1977, 1978).

Most of these carcinogenic metals are positive in short-term assays for genetic damage. Hexavalent chromium is mutagenic in the Ames test (Petrilli and de Flora, 1977; 1978). Arsenic com-

TABLE 5.3

Mutagenicity of Chloroform-Methanol Extracts of Several Macroalgae from the Gower Peninsula, West Glamorgan, U.K., in *Salmonella typhimurium* TA100 and TA98 with (+S-9) or without (–S-9) Activation by Rat Liver Microsomes

		TA100		*TA98*	
		–S-9	+S-9	–S-9	+S-9
Enteromorpha sp.	25 μl	140.0±32.66	92.0±7.21	10.5±2.38	13.25±1.26
	12.5 μl	145.5±25.75	99.0±21.71	—	—
	7.5 μl	140.5±14.55	104.25±6.70	—	—
Porphyra umbilicalis	25 μl	127.5±5.07	125.5±9.95	6.5±3.70	12.25±4.92
	12.5 μl	108.25±12.15	143.5±14.55	—	—
	7.5 μl	106.5±13.08	136.75±23.43	—	—
Cystoclonium purpurem	25 μl	159.5±14.11	118.25±16.26	12.0±2.71	17.0±4.83
	12.5 μl	161.75±19.69	118.25±5.97	—	—
	7.5 μl	143.0±18.57	109.75±35.70	—	—
Plocamium coccineum	25 μl	873.0±149.18	434.5±58.86	21.75±1.26	35.5±4.04
	12.5 μl	885.0±110.98	339.75±120.13	—	—
	7.5 μl	692.0±137.91	283.0±26.86	—	—
DMSO	25 μl	137.25±20.21	90.5±11.96	16.5±5.07	23.25±7.54
Methyl methane-sulfonate	2 μl spot	+	—	—	—
Benzo(a)pyrene	5 μg	—	—	—	102.75±9.25

Notes: Concentrations of extract are given in μl/plate. Data are means and standard deviations for four replicate plates.

pounds are not generally mutagenic in bacteria, but are potent clastogens for mammalian cells in vitro (Leonard and Lauwerys, 1980). Cadmium, nickel, cobalt, and lead also induce chromosome aberrations in cultured mammalian cells (Flessel, 1979). A complete discussion of the genetic activity of carcinogenic heavy metals is given by Sunderman (1979) and Flessel (1979). It should be appreciated that the salt form or oxidation state of the heavy metal is important in determining its genetic activity. A negative result from an algal extract should be evaluated in the light of this fact before the absence of a heavy metal carcinogen is accepted.

Studies by Bryan and Hummerstone (1973) and Haug, Melsom, and Omang (1974) have established a relationship between the concentrations of heavy metal found in seawater and the amounts accumulated by *F. vesiculosus* and *A. nodosum*. Another study has demonstrated this relationship using *Enteromorpha* spp. and *U. lactuca* (Seeliger and Edwards, 1977). However, there are a number of variables besides ambient heavy metal concentration that may affect the levels present in algae. Among these are seasonality, age of the plant, toxicity, position on the shoreline, salinity, water temperature, and possible interactions of the metals during the uptake process (Phillips, 1977). For this reason it may be better to regard algal bioassays as a semiquantitative technique, but this does not reduce their value as inexpensive, integrative prescreens for identifying polluted areas. The concentrations of suspect heavy metal carcinogens that have been measured in marine algae are given in Table 5.4. These values generally represent concentration factors over ambient water levels of 10^2 to 10^4.

TABLE 5.4
Concentrations of Carcinogenic or Potentially Carcinogenic Heavy Metals in Marine Macroalgae

	Cd	*Pb*	*Ni*	*Co*	*Cr*	*As*
Fucus vesiculosus	0.05-25.6	0.5-202	1.2-29.6	5.5-11.3	3.8-4.5	35.2-35.8
Fucus serratus	2.3-13.0	4.54	—	—	—	—
Ascophyllum nodosum	0.7-16.0	0.7-95	3.9-6.3	—	2.2-2.8	—
Enteromorpha spp.	0.7-13.0	6-1,200	—	—	—	—
Ulva lactuca	0.5-4.8	10-66	8-33	4-40	—	—
Porphyra umbilicalis	0.05-21.0	0.8-10.5	0.2-9.7	—	—	—

Note: Concentrations are in ppm of dry tissue.

Source: Phillips (1977).

Organic Compounds

There is little information available on the concentration of organic compounds by marine macroalgae (Phillips, 1978), but a few reports have been published. Robinson et al. (1967) measured the concentrations of organochlorine residues in a variety of species from different trophic levels of a marine ecosystem. It was found that *F. serratus* accumulated 0.001 ppm of HEOD (an oxygenated class of organochlorine pesticides) and 0.002 ppm of p,p′-DDE; no concentration factors over ambient levels in the sea were given. These values were approximately an order of magnitude lower than those measured in *M. edulis* and *Cardium edule*, which may further concentrate organochlorines contained in the phytoplankton upon which they feed. Parker and Wilson (1975) analyzed *Pelvetia*, *Fucus*, and *Ascophyllum* for unspecified polychlorinated biphenyls (PCBs) in the vicinity of a sewage outfall known to be discharging contaminated sludge. PCB concentrations in *Fucus* and *Pelvetia* showed the same trends that had previously been observed in mussels in the same estuary. Levels of PCB in *Ascophyllum* did not correspond to these trends, and showed no variation with distance from the discharge site. Concentrations as high as 0.34 and 0.31 ppm wet weight were measured for *Pelvetia* and *Fucus*, respectively.

Preliminary studies have been conducted by Levine and Wilce (1980) in *U. lactuca* at 28 sites along the Atlantic coast from New Hampshire to Rhode Island. At various sites elevated levels of PCBs, the organochlorine pesticides aldrin, dieldrin, DDE, and DDT, and the organophosphate pesticides diazinon, malathion, and parathion have been found.

The concentration of polycyclic aromatic hydrocarbons (PAH) by *Fucus* spp. has been investigated by Dunn (1980). A high correlation was found between concentrations of benzo(a)pyrene (B(a)P) in *Fucus* spp. and *M. edulis* from the same locations. Levels of B(a)P in *Fucus* were approximately one-half those found in *Mytilus*. It was also found that concentrations of benz(a)anthracene, benzo(e)pyrene, benzo(b)fluoranthene, benzo(a)pyrene, chrysene, and coronene in *Fucus* were the same as in associated sediments, suggesting that a dynamic equilibrium may exist between PAH in the sediments, water column, and seaweed. Lower-molecular-weight PAH (fluoranthene and phenanthrene) are not correlated with concentrations of B(a)P. It is suggested by the author that this is because they are much more hydrophilic, and hence less likely to partition out of the water.

A preliminary study of marine macroalgae as bioconcentrators of mutagenic pollutants was conducted by Barnes and Parry (unpublished data), using *F. vesiculosus* collected at different sites along the Gower Peninsula in Wales. The initial plan was to develop a bioassay for heavy metal mutagens or carcinogens; hence standard procedures were used to make tissue digests in nitric-perchloric acid. These were neutralized and tested for mutagenicity using the microtitre fluctuation test. In this test an indicator strain of bacteria (*S. typhimurium* TA98) is added to selective growth medium with the extract to be tested. The mixture is then dispersed into the 96 wells of a microtitre plate and incubated for three to four days. During this time only mutant cells can grow and divide. The number of wells in which growth has occurred at the end of incubation is therefore an indication of the mutagenicity of the substance under test.

The results for digests of *Fucus* are shown in Figure 5.1. The abscissa indicates the distance from the industrial center of Swansea-Port Talbot and the location of the collection sites. The bars indicate the mutagenic activity of each extract when tested at 0.5, 0.25, and 0.1 ml/plate. Digests of plants collected close to the industrial center of Swansea-Port Talbot were clearly more mutagenic than digests from plants growing farther away. It can be shown that these results are not attributable to artifacts such as nutrilite supplementation, biosynthesized mutagens, or the formation during digestion of

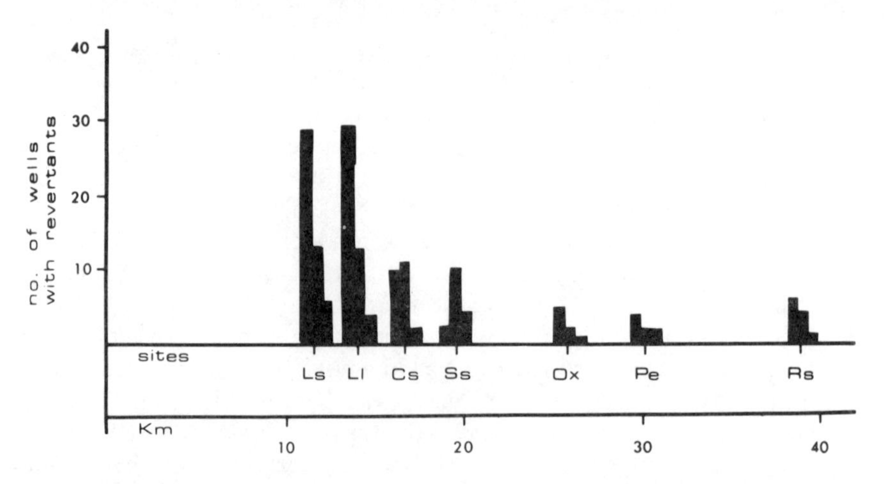

FIGURE 5.1. Mutagenicity of Nitric-perchloric Acid Digests of *Fucus vesiculosus* (*F. serratus* at Pe) from Sites on the Gower Peninsula, Wales, U.K., for *Salmonella typhimurium* TA98 in the Microtitre Fluctuation Test.

mutagenic N-nitrosamines. It may be that these are organic rather than heavy metal mutagens. Similar studies with digests of *Enteromorpha* spp. have indicated the presence of two distinct activities, both of which exhibit heat-deactivation kinetics that would be expected of organics.

Techniques Used in the Algal Macrophyte Bioassay

Collection Techniques

In designing a study that uses biological indicators, it is imperative that potential sources of variation be recognized and controlled. Experimental and control sites should be carefully matched for environmental parameters, and prescribed collection methods should be adhered to.

One potential source of variation in long-lived algae is the age-related difference in contamination levels exhibited by different parts of the plant. This has been observed for zinc, lead, copper, and iron in *Fucus*, and it is hypothesized that younger portions of the thallus may require some time to attain equilibrium with the surrounding water. Season is probably also important, since rates of growth vary throughout the year; an increasing concentration gradient from new tissue to old would be less evident at times when less new growth is produced. It is clear that similar portions of the algal thallus should be sampled at all sites.

It is also important to minimize ecological differences, since environmentally induced changes in phenotype are possible. In *Laminaria*, for example, it is known that alginate content may vary considerably, and that most of the variance can be partitioned statistically into a nongenetic component (Chapman and Doyle, 1978). Some or all of this nongenetic variance may be environmental. Since variation in alginate content might affect the permeability of the cell wall to mutagens, lack of attention to details of site selection could result in spurious conclusions.

Although no information is available on variation of lipid levels in macroalgae, this has proved to be an important factor influencing accumulation of organochlorines in bivalves (Phillips, 1978), and should be borne in mind as a possible confounding factor.

Physical factors may markedly affect the solubility of heavy metals and organochlorines, and presumably of other organics as well. Data on this point are sketchy, but Biggar, Doneen, and Riggs (1966) have shown an increase in solubility with temperature for

DDT, lindane, aldrin, dieldrin, and endrin. This must be regarded as a general chemical principle relating to all compounds. Higher salinities generally result in decreased solubility of heavy metals and an increase in the proportion precipitated or adsorbed to particulates (Phillips, 1977), and this probably is true for nonpolar organic compounds as well.

The concentration of heavy metals in macroalgae has been shown to vary with the position of the plant on the shoreline (Nickless, Stenner, and Terrille, 1972; Fuge and James, 1974; Bryan and Hummerstone, 1973), and may be due to differences in exposure to waterborne contaminants during the tidal cycle. An additional subtlety is the occurrence of stratification between freshwater and saltwater zones in estuaries, which may vary in depth and in the amount of intermixture. Selective partitioning of pollutants into one zone or the other may occur. In addition, nonpolar compounds (which include many mutagens and carcinogens) tend to be concentrated near the surface rather than evenly distributed through the water column (Longwell, 1977).

Extraction and Fractionation Techniques

This topic has been the subject of a review by Epler (1980), so a lengthy discussion is not necessary here. However, a brief description of the chemical techniques available and a few individual comments may be useful. It should be emphasized that there is no general procedure for dealing with complex substances; in fact, successful methodologies are likely to be modified according to the chemical classes under study. In an environmental study where the identity of potential mutagens is unknown, it may be prudent to use several variations.

Extraction in organic solvents is the most flexible method for preparing a crude extract, since a number of techniques and solvents are available. Tissue may be homogenized with solvent in a blender or Potter-Elvehjem apparatus, Soxhlet-extracted, or simply extracted in solvent for various periods of time. A number of solvents have been used for extraction, including DMSO, ethanol, chloroform or chloroform-methanol mixtures, ethyl ether, hexane, cyclohexane, and methylene chloride. In general, solvents of intermediate polarity will be most useful when working with unknowns, since they will extract the broadest range of compounds. Some solvents can give rise to artifactual mutagenic activity. Ethyl ether may decompose into mutagenic peroxides, and may also leave behind highly toxic residues after evaporation (Durston and Ames, 1974). Methylene chloride can be mutagenic in the Ames *Salmonella*/mammalian microsome system (Epler, 1980). Another common procedure involves precipitation with ammonium sulfate and/or basification with ammonium hydroxide prior to organic extraction, but if compounds with carbonyl groups are present, this practice may result in a chemical reaction (the Maillard browning reaction) that generates mutagenic species (Iwaoka et al., 1981).

Crude extracts of untreated or drinking water samples have been prepared by column chromatography using Amberlite XAD-2 or XAD-4 resin. Aqueous samples are first passed through the column, then organics are eluted with acetone or a similar solvent (Glatz et al., 1978; Nestmann et al., 1979; Cheh et al., 1980). This technique would not be suitable for tissue homogenates, although Baylis, Sparks, and Chang (1980) have used XAD-2 for removing histidine from organic extracts prior to bioassay in the Ames test. Reconstruction experiments have shown, however, that the resin does not efficiently bind some carcinogens, such as benzo(a)pyrene (Yamasaki and Ames, 1977).

A number of techniques are available for separating crude extracts into fractions of chemically similar compounds. Epler et al. (1978) used a liquid extraction scheme for separating six acidic, basic, or neutral fractions. Another preparative scale technique is column chromatography using florisil, alumina, or silica. Reversed-phase column packings are also available, and may present a useful alternative when the reactivity of silica gel is a problem. Preparative scale thin-layer chromatography can also be used. High-pressure liquid chromatography is an extremely powerful technique, but is expensive and specialized.

Testing Extracts and Fractions for Genetic Activity

A number of microbial assays are available for testing the genetic activity of complex mixtures. The standard Ames plate incorporation test uses specially constructed strains of *S. typhimurium* auxotrophic for histidine. Mutagenicity is assayed by plating a bacterial suspension and the test compound on agar plates lacking histidine. Colonies growing on these plates after incubation must do so as a result of mutation to histidine independence. The test is sometimes modified by introduction of a preincubation step. Details of the methodology can be found in Ames, McCann, and Yamasaki (1975) and Yahagi et al. (1975). The Ames strains, and others, can also be used in a modified fluctuation test (Green, Muriel, and Bridges, 1976; Gatehouse, 1978; Gatehouse and Delow, 1979). *Saccharomyces cere-*

visiae D7 is a strain of yeast constructed to measure simultaneously the induction of mitotic recombination, gene conversion, and mutation (Zimmermann, Kern, and Rasenberger, 1975). The differential sensitivity of DNA repair-deficient and repair-proficient bacterial strains to the toxic effect of a chemical can be used as an indicator of DNA damage. Chromosomal aberrations in mammalian cells can also be measured on a routine basis. The essential point in this discussion is that a bioconcentrator extract can be quickly tested for activity at a number of genetic end points, and this reduces the chance of drawing erroneous conclusions from any single test.

It must be appreciated that testing complex mixtures involves additional factors that need not be considered in the design and interpretation of experiments with pure compounds. For example, biological extracts that are found to be mutagenic for an auxotrophic bacterial strain must also be assayed for nutrilite contamination. This is necessary to exclude the possibility of uncontrolled supplementation and the increased number of spontaneous revertants that would result from an increased number of cells per plate. False negative results are also possible. The inclusion of a number of toxic substances in the mixture may preclude detection of a mutagen by making it impossible to assay the active dose range. Another concern is that mutagens may partition into microdroplets of hydrophobic molecules, and therefore not be available for uptake by the tester organism. Such a situation, involving the suppression of mutagenic activity of benzo(a)pyrene by crude oil, has been demonstrated by Petrilli, de Renzi, and De Flora (1980).

It is possible to remedy these misleading situations by fractionating the crude extract, as mentioned above. On the other hand, genetic activity may be detected in a crude extract as a result of additivity among a number of weak mutagens (Kaden, Hites, and Thilly, 1979). Fractionation of such an extract will separate these components, and the activity of each individual compound will then fall below the limits of detection for the assay. These procedures cannot, therefore, be applied uncritically.

Mutagenic synergism (comutagenicity, antimutagenicity, desmutagenicity, metabolic potentiation) leads to a further level of complexity. The mutagenic activity of some chemicals can be enhanced by others, either through interaction with the mutagen itself (Stich and Kuhnlein, 1979; Stich, Wei, and Whiting, 1979) or by interaction with the enzymatic activation pathway (Nagao et al., 1977; Fujino et al., 1978; Ashby and Styles, 1978; Rao et al., 1979; Sugimura and Nagao, 1980). The same two types of interaction may also inhibit the action of some mutagens (Levitt and Neibert, 1977; Lo and Stich, 1978; Kada, Morita, and Inoue, 1978; Stoltz et al., 1979; Rosin and Stich, 1979; 1980; Lai, Butler, and Matney, 1980). Fractionation could be expected, in many cases, to separate the mutagen and the synergist, and thus to increase or decrease the mutagenic activity of the extract. In extreme cases fractionation may even reverse the results of an initial experiment; activity may disappear altogether, or hitherto undetected activity may appear. The interpretation to be placed on such data is still a matter of indecision (Ashby and Styles, 1978; Rosin and Stich, 1980), but it is clearly important to recognize these complexities when they occur.

UNEXPLOITED APPROACHES USING ALGAL MACROPHYTES AS AN IN SITU ASSAY FOR GENETIC ACTIVITY

Another approach to the detection of dispersed mutagens in the marine environment is the use of an in situ bioindicator in which damage to the genetic system is manifested by some outward change in phenotype. Genetic activity can thus be measured directly, and the ambiguities inherent in extraction procedures and in vitro assays are avoided. This approach would be useful in detecting labile mutagens that may not have a long lifetime in the environment. It also offers a way of using organisms that activate and detoxify carcinogens, and therefore cannot be used in conjunction with in vitro assays, although they may otherwise be good indicators.

Little, if anything, has been done in this area, despite the fact that much is known about the genetics of various marine macrophytic algae or, at the least, the techniques are in place that are required to gain this information. It is therefore relevant to summarize what is known and to show how it might be applied to environmental mutagenesis. In this section no attempt will be made to propose finished or detailed protocols; these are elements of the research that needs to be done. Rather, salient points about the life cycles and genetics of several algae will be briefly reviewed, and strategies for exploiting these characteristics to derive information on genetic damage will be considered.

Ulva spp.

A distinct advantage in using *Ulva* is that drastic changes in morphology can be identified, yet the level of organization is low enough that such mutants remain fully viable and fertile. Another, complementary advantage is that all cells of the thallus can be induced to form gametes, either directly (from the blade) or indirectly (by regeneration from the holdfast). Genetic analysis of a mutational chimera is therefore possible, in contrast with higher plants in which reproduction is carried out by specialized organs and genetic analysis is possible only when these reproductive organs arise in the mutant sector. A caveat is in order, however. All work in *Ulva* has been done on *U. mutabilis*, a European species. Although it is reasonable to suppose that the results of this research could be applied to other common species, such as *U. lactuca*, this has not yet been done.

A disadvantage is that changes in pigmentation, an easily visible class of mutation that has been widely exploited in higher plants and other algae, may be of limited usefulness in *Ulva*. This is because such mutations, which tend to be lethals, would be expressed in the vegetative haploid gametophyte and eliminated. As a general rule, genetic analysis of any lethal in *Ulva* is therefore not possible. A corollary of this is that genetic load in populations of *Ulva* might be expected to be negligible.

The life cycle of *Ulva* is diplobiontic (having both gametophytic and sporophytic plants) and isomorphic (both stages are morphologically similar). The life cycle is shown in Figure 5.2. The diploid sporophyte produces, through meiosis, quadriflagellate zoospores. After a brief period of activity, these settle on the substrate and germinate into gametophytes. Gametophytes are either + or - in mating type, and the gametes that they produce fuse with a gamete of the opposite mating type to produce a diploid sporophyte. The

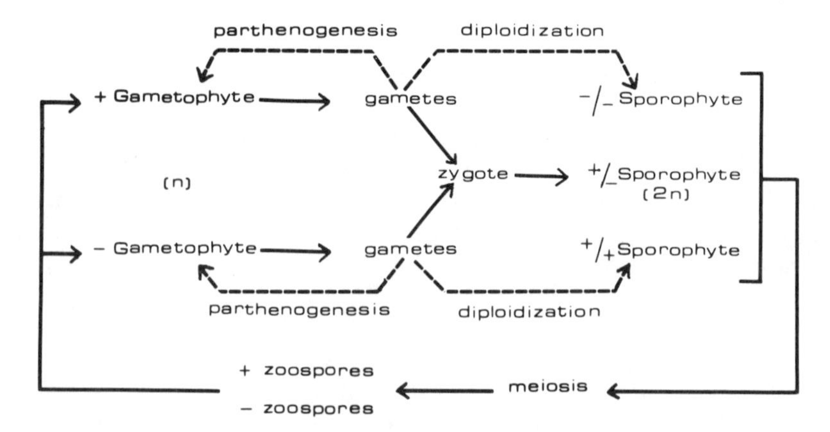

FIGURE 5.2. Life Cycle of *Ulva mutabilis*.

life cycle is complicated by the fact that gametes may undergo parthenogenetic development, either immediately into gametophytes or, after diploidization, into sporophytes. The life cycle of *U. mutabilis* is discussed in further detail by Løvlie (1964) and Fjeld and Løvlie (1976).

A basic protocol for screening genetic damage in *Ulva* might involve the following basic steps:

1. Collection of plants from polluted and clean sites. For reasons that have already been discussed, experimental and control populations should be matched as closely as possible to minimize environmentally induced changes in phenotype.
2. Identification of plants as sporophytes or gametophytes. Plants can be stored at 5°C under low light (Fjeld, 1972) while fragments of thallus are tested for the production of biflagellate gametes or quadriflagellate zoospores. Plants proving to be gametophytes are discarded.
3. Induction of zoosporogenesis in sporophytes, and germination of the resulting zoospores. The way in which this part of the experiment is arranged depends on the mutation frequency per cell. It would be desirable to have several mutational events per culture dish, and this would determine the size of the thallus fragment used. For *U. mutabilis*, where a forward mutation frequency (over all loci) on the order of 10^{-4} has been observed (Fjeld and Børreson, 1975), about 10^{4} cells should probably be used. Another technical consideration is that enough room must be present in the culture vessel for the germination of all zoospores without the occurrence of selection. Mutation frequency therefore will also determine the type of culture vessel.
4. Screening at periodic intervals for the expression of new mutations in the young haploid gametophytes. The time at which scoring is done depends on the mutations that are expressed and the developmental program of the plant. In *Ulva* this should probably be weekly (see below for discussion of mutations in *Ulva*).

There are at least two classes of mutation that might be used to assess genetic damage. Monophasic or polyphasic lethals could be scored at an early stage after germination of the zoospores, while data on developmental or morphological mutants could be obtained from the same experiment after two to three weeks.

Monophasic or polyphasic lethals are alleles that are expressed only at one time or several times during the life cycle. If they are induced in a cell after the developmental event that they control has already occurred, cell death will not be immediate. Such mutations might occur in both the gametophyte and the sporophyte of *Ulva*. If they are also recessive, they would never be expressed in the diploid sporophyte. A reasonable strategy is to screen for the occurrence of these mutations in individual cells of the sporophyte by examining the phenotypes of the haploid gametophytes that descend from them. After one week the germling gametophyte contains several cells, and between two and three weeks it reaches the 20-30-cell stage (Løvlie, 1964). Gametophytes dead or moribund at these times are presumed to carry a lethal. Only germlings that have divided at least once should be scored, to avoid the objection that such lethality is due exclusively to toxic action upon the genetic or metabolic machinery of the parent cell, and has been transferred to the zoospore via a type of maternal effect.

Unfortunately, as mentioned above, it is not possible to show unambiguously that such phenotypes are true mutants, since genetic analysis of dead plants is proscribed. As a technical aside, it is probably a better practice to score for the frequency of mutant events rather than of mutants, since a certain degree of random elimination is to be assumed for the latter. Since each sporangium (cell) in *Ulva* produces four to eight zooids, depending on species, a sporangium in which a lethal is segregating should produce one-half that number of gametophytes expressing the mutant phenotype, and scoring should be done in multiples of two or four. The great advantage of using lethals as an end point for genetic damage is that it is a multilocus phenotype. The mutant frequency, and hence the sensitivity, of the assay increases in proportion to the number of loci involved.

Developmental mutants have been studied in detail by a number of workers, and this work is reviewed by Fjeld and Løvlie (1976). In brief, a number of mutants are known to lead to aberrations in the final morphology of the plant. Slender mutants do not develop the normal flattened blade, but retain a ribbonlike form. Precocious mutants begin to develop a flattened thallus earlier than usual, and lack a proper holdfast. When such mutations are considered, it should be noted that in bubble, another developmental mutant, there is a strong predetermining effect that causes 20-80 percent of genotypically *bu* gametophytes to resemble phenotypically the wild-type spore mother cell from which they originate. The proportion of wild-type phenocopies is influenced by culture conditions such as lighting (Fjeld, 1972). It is therefore important, when screening developmental mutants, to ensure that their behavior is fully understood and that all sources of error are controlled. Another point that merits discussion is the mutational instabilities that have been observed in *U. mutabilis*. Reversion of slenderlike and bubble mutants to wild type is much more frequent than the original forward mutation. Mutation of these revertants to slenderlike again is also much higher, leading to the hypothesis that an instability of some kind arises at this locus (Fjeld and Børreson, 1975). Among the progeny of wild-type plants that have not previously mutated, there is no predominance of one mutation over another. Mutational instabilities would therefore not be expected to cause a problem in plants freshly collected from the wild.

The techniques needed for genetic analysis of *Ulva* are described in the literature. Løvlie (1964) explains the developmental sequence. Culturing techniques are covered by Fjeld and Løvlie (1976) and Nordby and Hoxmark (1972). Fjeld (1972) discusses the induction of zoosporogenesis and the establishment of mass gametophyte cultures.

Phaeophyta

Much less work has been done with the brown algae, although genetical studies have been performed. For the most part these have been related either to cross-breeding experiments for the purpose of studying systematic relationships or to studies of alginate production and heritability with a view to commercial application. As a result, little is known about the mutations that may occur in common brown macroalgae such as *Laminaria*, *Fucus*, and *Ascophyllum*. On the other hand, the techniques needed for genetical analysis are well established, and should be easily applicable to assays for genetic damage in these common genera.

The life cycle of *Fucus* spp. is shown in Figure 5.3(a), and the life cycle of *Laminaria* spp. in Figure 5.3(b). The life cycle of *Ascophyllum* spp. is identical to that of *Fucus*. *Fucus* has a haplobiontic life cycle, the dominant phase being diploid. The swollen tips, or receptacles, are fertile. Gametes are produced in oogonia and antheridia, which are borne in conceptacles, or cavities within the receptacle. Gametes are shed into the open ocean, where fertilization occurs, followed by the development of the zygote into a diploid plant. In *Laminaria* the life cycle is diplobiontic, with both a diploid sporophyte and a haploid gametophyte stage. It is fundamentally similar to that of *Ulva*, except that the gametophyte and sporophyte differ markedly in size, structure, and longevity. The *Laminaria* gametophyte is microscopic, branched, and ephemeral, whereas the sporophyte is large, complex, and perennial.

Fucus

The life cycle of *Fucus* suggests that techniques that have been used to detect the presence of recessive lethals in conifers could be

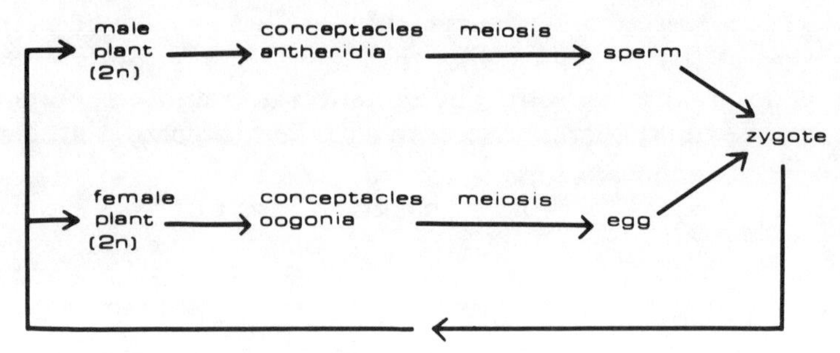

FIGURE 5.3 (a) Life Cycle of *Fucus vesiculosus.*

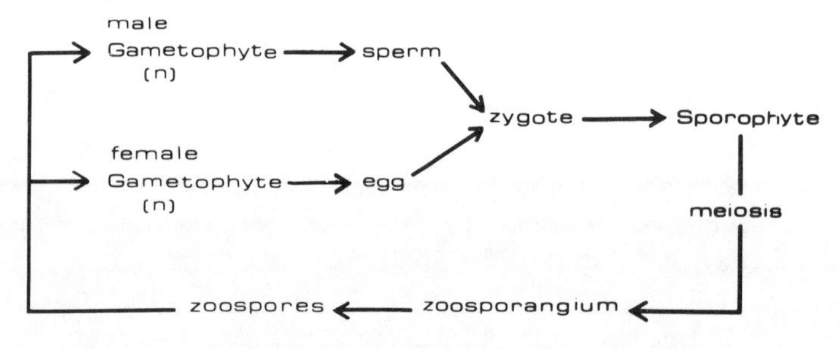

(b) Life Cycle of *Laminaria* spp.

used equally well on this alga. Sorenson (1969) and Franklin (1972) have obtained estimates of the number of lethal equivalents per diploid genome by measuring the ratio of viable zygotes produced in self-fertilizations and cross-fertilizations of an individual tree. The following relation is then applied:

$$2B = -4 \ln (\text{viable selfed zygotes}/\text{viable out-crossed zygotes})$$

where B is the number of lethal equivalents per genome. A lethal equivalent may be either a single, fully lethal gene, or a number of sublethal genes whose effects will sum to produce a lethal phenotype.

This technique requires the use of monoecious species, because of the requirement for selfing. Unfortunately, *F. serratus* and *F. vesiculosus* are dioecious, but *F. spiralis* and *F. distichus* are monoecious.

The general sequence of steps that would be required for an assay of this kind with *Fucus* are as follows:

1. Collect fertile plants from matched experimental and control sites, observing the precautions detailed above. Store plants until needed.

2. Induce gametogenesis in the receptacles of a matched experimental and control plant.
3. Separate the sperm and eggs by filtration. Sperm can be inactivated to prevent uncontrolled fertilization. Adjust titre to a standardized value.
4. Perform the self- and cross-fertilizations by mixing appropriate eggs and sperm.
5. After six hours, pipette off unfertilized eggs and excess sperm (zygotes will attach to bottom of dish). Decontaminate the zygotes with a rinse of penicillin-streptomycin or chloramphenicol.
6. Culture the germlings. Score for percentage of viable progeny two weeks after fertilization, and calculate the lethal equivalents per genome.

The assumption in applying this technique is that new lethals are not induced during vegetative growth and gametogenesis. This is not likely to be true, and careful interpretation of the results is necessary. Klekowski (1978) has made the point that the morphology and developmental pattern of a plant will determine the fate of an induced mutation. In *Fucus*, for example, the detection of a lethal mutation requires that most or all of the gametes of a given receptacle carry it. This can occur only when the mutation arises in the apical cell and is passed to the seven or eight rapidly dividing initials that derive from it. Mutations that originate in individual initials or other somatic cells may be passed to some gametes, but not in sufficient numbers to allow detection. Moreover, the time at which new mutations become established in the apex is important. *Fucus* grows by dichotomous branching; if the new mutation is established early, it will be carried in the apices of many growing tips. If this event occurs at progressively later times, the mutation will be found in correspondingly fewer tips (Figure 5.4[a]).

This consideration determines the type of data that can be expected, and also suggests some features of an effective sampling design. Plants from a clean environment that are not subject to the induction of new lethal mutations during vegetation growth should yield the same estimates of lethal equivalents per genome when receptacles from widely separated parts of the plant body are analyzed. If lethal mutations have been induced and stabilized in some apices, this will not be true. The estimates of lethal equivalents per genome derived from different receptacles will be distributed with a larger standard deviation, skewed toward the high end, or may have a binodal distribution (Figure 5.4[b]). Alternatively, any of these distributions might occur with an elevated median value. This might be the result either of a higher population frequency of lethal alleles,

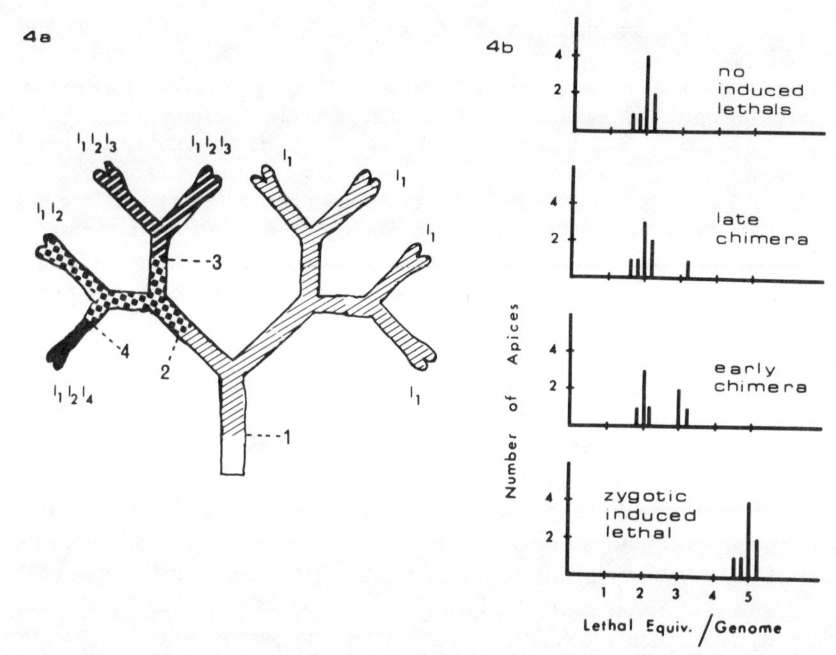

FIGURE 5.4. (a) The chimeric distribution of mutations (1–4) occurring at successively later times during growth of the thallus. (b) Representation of lethal equivalent data that would be expected as a consequence of multiple chimeras.

or of the induction of new lethals in the zygote immediately after fertilization. If there were a consistent difference in this regard between plants from clean and polluted populations, such data might be useful. Unless a large number of plants were analyzed, however, interplant variation could result in an overlap of the two distributions that would not be statistically distinguished. Analysis of intraplant variation, which eliminates this source of error, is therefore likely to be a more efficient means of detecting an increased induction of mutations.

It should be recognized that this technique is essentially different from those in which mutants are scored directly and their frequency is expressed in terms of a total population size. The parameter estimated in this case is the induction of mutations in an apical cell during a given period of growth, and the common denominator is the number of loci in the genome rather than a population of cells or germlings. This means that the sensitivity of the assay depends primarily on the size and mutability of the *Fucus* genome. The number of apices and plants analyzed can be quite small, although the exact number required to generate adequate statistical data would need to be determined by experiments.

Methods for the genetic manipulation of *Fucus* are discussed by Pollock (1969; 1970); McLachlan, Chen, and Edelstein (1971); and Quatrano (1974). The primary reproductive period in *Fucus* is in winter and spring; mature receptacles can be found in *F. distichus* from December to June (McLachlan, Chen, and Edelstein, 1971). Plants can be stored for two to three weeks in a refrigerator at 4°C (Quatrano, 1974). Media and culturing methods are described by McLachlan, Chen, and Edelstein (1971). Methods for inducing the release of gametes are given in Quatrano (1974) and Pollock (1970). A principal difficulty in performing controlled crosses in monoecious species, where both male and female gametes may be released together, is the separation of eggs and sperm. This can be accomplished by inactivating the sperm in 10^{-3}M EDTA or 1 percent chloral hydrate (Quatrano, 1974), or by filtering through Nitex nylon mesh (Quatrano, 1974; Pollock, 1969). A ratio of approximately 500 sperm per egg is optimal (Bolwell et al., 1977). Fertilized eggs will settle on a glass or plastic dish within four to six hours; unfertilized eggs do not, and can be removed by pipetting along with any excess sperm (Quatrano, 1974). Zygotes may be visualized with Calcoflour White ST under ultraviolet microfluorescence (Bolwell et al., 1977). Zygotes may be washed with a solution of penicillin-streptomycin or chloramphenicol to prevent contamination in the cultures of young germlings (Quatrano, 1974).

Laminaria

Since the life cycle of *Laminaria* is similar to that of *Ulva*, similar methods of assay are possible. Indeed, screening for lethals, which are expressed shortly after germination of the zoospore, should be much easier due to the small size of the gametophyte. It has been reported that "several hundred" can grow on the surface of a cover slip (Lüning and Dring, 1972).

Morphological mutants have not been studied in *Laminaria* to any extent. In contrast with the green algae, pigmentation mutants could prove to be very useful. A mutational change in the accessory pigments that determine frond color would not necessarily block photosynthesis, and hence would not be lethal. An elegant parallel has been examined by van der Meer and his co-workers in genetic studies of *Gracilaria tikvahiae*, a red alga. Mutations that alter the red phycobiliproteins may change the color of the alga dramatically, but hardly affect viability (van der Meer and Bird, 1977; van der Meer, 1979). There are at least 30 cistrons involved in these changes, so that the frequency of color mutations should be quite high. Similar mutations probably also occur in the genes coding for the xanthophylls of *Laminaria*. These color mutations would be easily detected in the haploid gametophytes, and could be conveniently screened in conjunction with assays for lethals.

The parameter of interest remains the frequency of mutational events rather than the frequency of mutants themselves, and therefore, as with *Ulva*, attention must be paid to the scoring design. Sporangia contain 32-64 zoospores, depending on the species of *Laminaria*, and scoring must be done in units of this size. Methods for the manipulation and culture of *Laminaria* are given by Chapman (1973).

Mitotic Crossing-over

Studies of mitotic crossing-over in the soybean (*Glycine max*) have been described by Vig (1975) (see Chapter 2). Dark green or yellow sectors can be seen individually, or as twin spots on the light green background of the leaf. This is presumed to result from the operation of two codominant alleles that produce either a dark green or a yellow phenotype when homozygous, and a light green phenotype in the heterozygous state. Dark green or yellow spots in a heterozygous leaf may result from gene mutation, small deletions, loss of a chromosome segment, gene conversion, or nondisjunction. Twin spots arise from mitotic crossing-over.

The frequency of single and twin spots increases in a dose-related manner with exposure to chemicals such as caffeine, mitomycin C, sodium azide, methyl methanesulfonate, and ethyl methanesulfonate (Vig, 1975; Nilan and Vig, 1976). A provocative finding is that dimethylnitrosamine also induces an increase in the frequency of single yellow spots (Arenaz and Vig, 1978). This suggests that the soybean is capable of activating some promutagens, although it is not known how closely these products resemble those produced by mammalian cells in vivo.

It may be possible to develop algal analogues to this higher plant system. Although mitotic recombination has not been well studied in algae, it has been reported in *Chlamydomonas reinhardi* (Martinek, Ebersold, and Nakamura, 1970) and in *Gracilaria* sp. (van der Meer and Todd, 1977). The first requirement is for appropriate mutants that will produce distinctly different phenotypes in the heterozygote and each homozygote. As discussed above, such mutants are readily isolated in the red alga *G. tikvahiae* (van der Meer, 1979), and are potentially available in any alga in which color is determined primarily by accessory pigments rather than chlorophyll. No experiments have been done with *Gracilaria* to determine whether genotoxic agents increase the frequency of twin or single spots; however, it is probable that the chemicals that induce this kind of genetic damage in yeast (*S. cerevisiae*) and the soybean would have a similar effect in an alga. It would appear from the work with *Gracilaria* that there is a relatively high spontaneous frequency of mitotic crossing-over (van der Meer and Todd, 1977). Finding a sufficient number of sectors with genetic damage for statistical purposes would therefore not entail an excessive amount of work.

One potential difficulty associated with this approach is that the proper heterozygous strain must be constructed. Because of long life cycles, suitable, genetically defined mutants cannot be obtained quickly. Even when obtained, the special genotypes must still be transplanted to the sites of interest. This introduces the problems of plant loss due to failure of establishment and destruction by various manmade and natural disasters.

Another consideration is that not all mitotic recombination events will be equally visible. They can be seen only when the two daughter cells continue to divide so that a small patch of each genotype is formed. Van der Meer and Todd (1977) mention that this occurs only in regions of active division where "meristematic amplification" can occur. The growth habit and developmental pattern of the individual alga will therefore be important in determining how prominent twin spots will be.

SUMMARY AND CONCLUSIONS

Mutagens and carcinogens are undoubtedly entering the marine environment by way of rivers and other land runoff. It is known that some carcinogens are metabolized (activated?) by certain fish species, and there are cases where pollution is associated with neoplastic disease in fish. Mussels (*Mytilus* spp.) and other shellfish accumulate carcinogenic PAH such as benzo(a)pyrene and benz(a)-anthracene. There is therefore concern about the long-term effects on sea life and the possibility of human exposure.

Macrophytic algae may be useful monitors of carcinogenic and mutagenic pollution. This information is important because algae are not only a major component of the ecosystem but also at the bottom of the food chain. Furthermore, algae may accumulate contaminants from the soluble phase of the water column, whereas filter feeders sample predominantly suspended particulate matter.

Macroalgae may be used either as bioconcentrators of mutagens or directly as indicators of mutagenicity. The advantages of testing extracts from bioconcentrators with in vitro assays for genetic activity are the ability to screen many genetic end points and the enhanced sensitivity of some of these specially constructed strains. On the other hand, synergism may create ambiguities. While this type of assay may be more biologically relevant than strictly chemical methods, data cannot be unequivocally extrapolated to predict genetic hazards under actual environmental conditions. In situ systems are most relevant to the problem of assessing the biological significance of mutagenic pollution, since genetic damage is measured directly in members of the algal community. It should be possible to develop methods whereby some of the most abundant algal species can be manipulated to obtain this information.

ACKNOWLEDGMENTS

I would like to acknowledge many fruitful discussions with Dr. Edward Klekowski, many of whose ideas are presented here in modified form. I also wish to thank Dr. James Parry for his encouragement and support during the performance of some of the work discussed in this paper.

REFERENCES

Ames, B., J. McCann, and E. Yamasaki. 1975. "Methods for Detecting Carcinogens and Mutagens with the *Salmonella*/Mammalian-Microsome Mutagenicity Test." *Mutat. Res.* 31:347-363.

Arenaz, P., and B. K. Vig. 1978. "Somatic Crossing over in *Glycine max* (L.) Merrill: Activation of Dimethylnitrosamine by Plant Seed and Comparison with Methylnitrosourea in Inducing Somatic Mosaicism." *Mutat. Res.* 52:367-380.

Ashby, J., and G. Styles. 1978. "Co-mutagenicity, Competitive Enzyme Substrates and *in Vitro* Carcinogenicity Assays." *Mutat. Res.* 54:105-112.

Barry, M., and P. Yevich. 1975. "Part III. Histopathological Studies." *Mar. Pollut. Bull.* 6:171-173.

Baylis, J., T. Sparks, and C. Chang. 1980. "Survey of Estuarine Species as Bioaccumulators of Mutagens." *Environ. Mutagenesis* 2:265.

Biggar, J., L. Doneen, and R. Riggs. 1966. *Soil Interaction with Organically Polluted Water.* Summary Report. Davis: Department of Water Science and Engineering, University of California.

Bolwell, G. P., J. Callow, M. Callow, and L. Evans. 1977. "Cross-Fertilization in Fucoid Seaweeds." *Nature* 268:626-627.

Borowitzka, M. 1972. "Intertidal Algal Species Diversity and the Effect of Pollution." *Aust. J. Mar. Freshwat. Res.* 23:73-84.

Brown, E., J. Hazdra, L. Keith, I. Greenspan, J. Kwapinski, and P. Beamer. 1973. "Frequency of Fish Tumors Found in a Polluted Watershed as Compared to Nonpolluted Canadian Waters." *Cancer Res.* 33:189-198.

Bryan, G., and L. Hummerstone. 1973. "Brown Seaweed as an Indicator of Heavy Metals in Estuaries in South-West England." *J. Mar. Biol. Assoc. U.K.* 53:705-720.

Chapman, A. 1973. "Methods for Macroscopic Algae." In *Handbook of Psychological Methods. Culture Methods and Growth Measurements*, ed. J. Stein, pp. 87-104. London: Cambridge University Press.

Chapman, A., and R. Doyle. 1978. "Genetic Analysis of Alginate Content in *Laminaria longioruris (Phaeophyceae)*." *Proc. Int. Seaweed Symp.* 9:125-132.

Cheh, A., J. Skochdopole, P. Koski, and L. Cole. 1980. "Nonvolatile Mutagens in Drinking Water: Production by Chlorination and Destruction by Sulfite." *Science* 207:90-92.

Cole, H. 1979. "Pollution of the Sea and Its Effects." *Proc. Roy. Soc. Lond.* B 205:17-30.

Dawe, C., D. Scarpelli, and S. Wellings, eds. 1976. *Tumors in Aquatic Animals.* Vol. 20 of *Progress in Experimental Tumor Research*, ed. C. Dawe, D. Scarpelli, and S. Wellings. Basel: S. Karger.

Dunn, B. 1980. "Polycyclic Aromatic Hydrocarbons in Marine Sediments, Bivalves, and Seaweeds: Analysis by High Pressure Liquid Chromatography." In *Polynuclear Aromatic Hydrocarbons: Chemistry and Biological Effects*, ed. A. Bjφrseth and A. Dennis, pp. 367-377. Columbus, Ohio: Battelle Press.

Dunn, B. and H. Stich. 1975. "The Use of Mussels in Estimating Benzo(α)-pyrene Contamination of the Marine Environment." *Proc. Soc. Exp. Biol. Med.* 150:49-51.

Durston, W., and B. Ames. 1974. "A Simple Method for the Detection of Mutagens in Urine: Studies with the Carcinogen 2-Acetylaminofluorene." *Proc. Natl. Acad. Sci. U.S.A.* 71:737-741.

Environmental Protection Agency. 1976. *An Ordering of the NIOSH Suspected Carcinogens List Based Only on Data Contained in the List.* PB-251-851. Washington, D.C.: National Technical Information Service.

Epler, J. 1980. "The Use of Short-Term Tests in the Isolation and Identification of Chemical Mutagens in Complex Mixtures." In *Chemical Mutagens. Principles and Methods for Their Detection*, Vol. 6, ed. F. de Serres and A. Hollaender, pp. 239-270. New York: Plenum.

Epler, J., J. Young, A. Hardigree, T. K. Rao, M. Guerin, I. B. Rubin, C. H. Ho, and B. Clark. 1978. "Analytical and Biological Analysis of Test Materials from the Synthetic Fuel Technologies. I. Mutagenicity of Crude Oils Determined by the *Salmonella typhimurium*/Microsomal Activation System." *Mutat. Res.* 57:265-276.

Falkner, S., S. Endin, Y. Östberg, A. Mattisson, O. Johansson, M. Sjöbeck, and R. Fange. 1976. "Tumor Pathology of the Hagfish *Myxime glutinosa* and the River Lamprey, *Lampetra fluviatalis.*" In *Tumors in Aquatic Animals.* Vol. 20 of *Progress in Experimental Tumor Research*, ed. C. Dawe, D. Scarpelli, and S. Wellings, pp. 217-250. Basel: S. Karger.

Fjeld, A. 1972. "Genetic Control of Cellular Differentiation in *Ulva mutabilis.* Gene Effects in Early Development." *Dev. Biol.* 28:326-343.

Fjeld, A., and S. Børeson. 1975. "The Spontaneous Mutability in *Ulva mutabilis.* Occurrence of Unstable Strains." *Norw. J. Bot.* 22:77-82.

Fjeld, A., and A. Løvlie. 1976. "Genetics of Multicellular Algae." In *The Genetics of Algae.* Vol. 12 of *Botanical Monographs*, ed. R. A. Lewin, pp. 219-235. Oxford: Blackwell Scientific Publications.

Flessel, C. P. 1979. "Metals as Mutagenic Initiators of Cancer." In *Trace Metals in Health and Disease*, ed. N. Kharasch, pp. 109-122. New York: Raven Press.

Fossato, V. U., C. Nasci, and F. Dolci. 1979. "3,4-Benzopyrene and Perylene in Mussels, *Mytilus sp.* from the Laguna Veneta, Northeast Italy." *Mar. Environ. Res.* 2:47-53.

Fossato, V. U., and E. Sivievo. 1974. "Oil Pollution Monitoring in the Lagoon of Venice Using the Mussel *Mytilus galloprovincialis.*" *Mar. Biol.* 25: 1-6.

Franklin, E. C. 1972. "Genetic Load in Loblolly Pine." *Am. Nat.* 106:262-265.

Fuge, R., and K. James. 1974. "Trace Metal Concentrations in *Fucus* from the Bristol Channel." *Mar. Pollut. Bull.* 5:9-12.

Fujino, T., H. Fujiki, M. Nagao, T. Yahagi, Y. Seino, and T. Sugimura. 1978. "The Effect of Norharman on the Metabolism of Benzo(a)pyrene by Rat-Liver Microsomes *in Vitro* in Relation to Its Enhancement of the Mutagenicity of Benzo(a)pyrene." *Mutat. Res.* 58:151-158.

Gatehouse, D. 1978. "Detection of Mutagenic Derivatives of Cyclophosphamide and a Variety of Other Mutagens in a 'Microtitre R' Fluctuation Test Without Microsomal Activation." *Mutat. Res.* 53:289-296.

Gatehouse, D., and G. Delow. 1979. "The Development of a 'Microtitre R' Fluctuation Test for the Detection of Indirect Mutagens and Its Use in the Evaluation of Mixed Enzyme Induction of the Liver." *Mutat. Res.* 60:239-252.

Gesamp. 1977. *Impact of Oil on the Marine Environment.* Reports and Studies no. 6, pp. 1-128. Rome: Food and Agriculture Organization of the United Nations.

Glatz, B., C. Chriswell, M. Arguello, H. Svec, J. Fritz, S. Grimm, and M. Thomson. 1978. "Examination of Drinking Water for Mutagenic Activity." *J. Am. Water Works Assoc.* 70:465-468.

Green, M., W. Muriel, and B. Bridges. 1976. "Use of a Simplified Fluctuation Test to Detect Low Levels of Mutagens." *Mutat. Res.* 38:33-42.

Haug, A., S. Melsom, and S. Omang. 1974. "Estimation of Heavy Pollution in Two Norwegian Fjord Areas by Analysis of the Brown Alga *Ascophyllum nodosum.*" *Environ. Pollut.* 7:179-192.

Hellenbrand, K. 1978. "Effect of Pulp Mill Effluent in Productivity of Seaweeds." *Proc. Int. Seaweed Symp.* 9:161-171.

Heslop-Harrison, J. 1978. "Summary and Perspectives." *Environ. Health Perspect.* 27:197-206.

Hirose, H. 1978. "Composition of Benthic Marine Algae in Relation to Pollution in the Seto Inland Sea, Japan." *Proc. Int. Seaweed Symp.* 9: 173-179.

Hites, R., and K. Biemann. 1972. "Water Pollution: Organic Compounds in the Charles River, Boston." *Science* 178:158-160.

Ishio, S., K. Kawabe, and T. Tomiyama. 1972a. "Algal Cancer and Its Causes—I. Carcinogenic Potencies of Water and Suspended Solids Discharged to the River Ohmuta." *Bull. Jap. Soc. Sci. Fish.* 38:17-24.

——. 1972b. "Algal Cancer and Its Causes—II. Separation of Carcinogenic Compounds from Sea Bottom Mud Polluted by Wastes of the Coal Chemical Industry." *Bull. Jap. Soc. Sci. Fish.* 38:571-576.

Iwaoka, W., C. Krone, J. Sullivan, E. Meaker, C. Johnson, and L. Miyasato. 1981. "A Source of Error in Mutagen Testing of Foods." *Cancer Letters* 11:225-230.

Kada, T., K. Morita, and T. Inoue. 1978. "Anti-Mutagenic Action of Vegetable Factor(s) on the Mutagenic Principle of Tryptophan Pyrolysate." *Mutat. Res.* 53:351-353.

Kaden, D., R. Hites, and W. Thilly. 1979. "Mutagenicity of Soot and Associated Polycyclic Aromatic Hydrocarbons to *Salmonella typhimurium.*" *Cancer Res.* 39:4152-4159.

Klekowski, E. 1978. "Detection of Mutational Damage in Fern Populations: An *in Situ* Bioassay for Mutagens in Aquatic Ecosystems." In *Chemical Mutagens. Principles and Methods for Their Detection*, Vol. 5, ed. A. Hollaender and F. de Serres, pp. 79-99. New York: Plenum.

Kraybill, H. 1976. "Distribution of Chemical Carcinogens in Aquatic Environments." In *Tumors in Aquatic Animals.* Vol. 20 of *Progress in Experimental Tumor Research*, ed. C. Dawe, D. Scarpelli, and S. Wellings, pp. 3-34. Basel: S. Karger.

Kraybill, H., C. Dawe, J. Harshbarger, and R. Tardiff. 1977. "Aquatic Pollutants and Biologic Effects with Emphasis on Neoplasia." *Ann. N.Y. Acad. Sci.* 298.

Lai, C. N., M. A. Butler, and T. Matney. 1980. "Antimutagenic Activities of Common Vegetables and Their Chlorophyll Contents." *Mutat. Res.* 77:245-250.

Leary, J., R. Kfir, J. Sims, and D. Fulbright. 1979. "The Mutagenicity of Natural Products from Marine Algae." *Mutat. Res.* 68:201-305.

Lee, R. 1977. "Accumulation and Turnover of Petroleum Hydrocarbons in Marine Organisms." In *Fate and Effects of Petroleum Hydrocarbons in Marine Organisms and Ecosystems*, ed. D. Wolfe, pp. 60-70. New York: Pergamon Press.

Leonard, A., and R. R. Lauwerys. 1980. "Carcinogenicity, Teratogenicity and Mutagenicity of Arsenic." *Mutat. Res.* 75:49-62.

Levine, H., and R. Wilce. 1980. *Ulva lactuca as a Bioindicator of Coastal Water Quality.* Office of Water Resources Research, Project no. A-112-Mass. Amherst: University of Massachusetts.

Levitt, R., and D. Neibert. 1977. "Effects of Harman and Norharman on the Mutagenicity and Binding to DNA of Benzo(a)pyrene Metabolites *in Vitro* and on Aryl Hydrocarbon Hydroxylase Induction in Cell Culture." *Biochem. Biophys. Res. Commun.* 70:1167-1175.

Ljunberg, O. 1976. "Epizootiological and Experimental Studies of Skin Tumors in Northern Pike (*Esox lucius* L.) in the Baltic Sea." In *Tumors in Aquatic Animals.* Vol. 20 of *Progress in Experimental Tumor Research,* ed. C. Dawe, D. Scarpelli, and S. Wellings, pp. 156-165. Basel: S. Karger.

Lo, L., and H. Stich. 1978. "The Use of Short-Term Tests to Measure the Preventive Action of Reducing Agents on Formation and Activation of Carcinogenic Nitroso Compounds." *Mutat. Res.* 57:57-67.

Longwell, A. C. 1977. "A Genetic Look at Fish Eggs and Oil." *Oceanus* 20: 46-58.

Løvlie, A. 1964. "Genetic Control of Division Rate and Morphogenesis in *Ulva mutabilis* Föyn." *C. R. Trav. Lab. Carlsberg* 34:77-168.

Lüning, K., and M. J. Dring. 1972. "Reproduction Induced by Blue Light in Female Gametophytes of *Laminaria saccharina.*" *Planta* (Berlin) 104: 252-256.

Malins, D. 1980. "Pollution of the Marine Environment." *Environ. Sci. Technol.* 14:32-37.

——. 1977. "Biotransformation of Petroleum Hydrocarbons in Marine Organisms Indigenous to the Arctic and Subarctic." In *Fate and Effects of Petroleum Hydrocarbons in Marine Organisms and Ecosystems,* ed. D. Wolfe, pp. 47-59. New York: Pergamon Press.

Martinek, G., W. Ebersold, and K. Nakamura. 1970. "Mitotic Recombination in *Chlamydomonas reinhardi.*" *Genetics* (Suppl.) 64:S41-S42.

McLachlan, J., L. Chen, and T. Edelstein. 1971. "The Culture of Four Species of *Fucus* Under Laboratory Conditions." *Can. J. Bot.* 49:1463-1469.

Nagao, M., T. Yahagi, T. Kawachi, T. Sugimura, T. Kosuge, K. Wakabayashi, S. Mizusaki, and T. Matsumoto. 1977. "Co-Mutagenic Action of Norharman and Harman." *Proc. Jap. Acad.* 53B:95-98.

Neff, J. 1979. *Polycyclic Aromatic Hydrocarbons in the Aquatic Environment.* London: Applied Science Publications.

Nelson, C., and R. Hites. 1980. "Aromatic Amines in and near the Buffalo River." *Environ. Sci. Technol.* 14:1147-1149.

Nestmann, E., G. LeBel, D. Williams, and D. Kowbel. 1979. "Mutagenicity of Organic Extracts from Canadian Drinking Water in the *Salmonella*/ Mammalian-Microsome Assay." *Environ. Mutagenesis* 1:337-345.

Nickless, G., R. Stenner, and N. Terrille. 1972. "Distribution of Cadmium, Lead and Zinc in the Bristol Channel." *Mar. Pollut. Bull.* 3:188-190.

Nilan, R. A., and B. K. Vig. 1976. "Plant Test System for Detection of Chemical Mutagens." In *Chemical Mutagens. Principles and Methods for Their Detection*, ed. A. Hollaender, pp. 143-170. New York: Plenum.

Nordby, Ø., and R. C. Hoxmark. 1972. "Changes in Cellular Parameters During Synchronous Growth in *Ulva mutabilis* Føyn." *Exp. Cell Res.* 75: 321-328.

Pancirov, R., and R. Brown. 1977. "Polynuclear Aromatic Hydrocarbons in Marine Tissues." *Environ. Sci. Technol.* 11:989-992.

Parker, J., and F. Wilson. 1975. "Incidence of Polychlorinated Biphenyls in Clyde Seaweed." *Mar. Pollut. Bull.* 6:46-47.

Petrilli, F., and S. de Flora. 1978. "Metabolic Activation of Hexavalent Chromium Mutagenicity." *Mutat. Res.* 54:139-147.

——. 1977. "Toxicity and Mutagenicity of Hexavalent Chromium on *Salmonella typhimurium.*" *Appl. Environ. Microbiol.* 33:805-809.

Petrilli, F., G. de Renzi, and S. de Flora. 1980. "Interaction Between Polycyclic Aromatic Hydrocarbons, Crude Oil and Oil Dispersants in the *Salmonella* Mutagenesis Assay." *Carcinogenesis* 1:51-56.

Phillips, D. 1979. "Trace Metals in the Common Mussel, *Mytilus edulis* (L.), and in the Alga *Fucus vesiculosus* (L.) from the Region of the Sound (Öresund)." *Environ. Pollut.* 18:31-43.

——. 1978. "Use of Biological Indicator Organisms to Quantitate Organochlorine Pollutants in Aquatic Environments—a Review." *Environ. Pollut.* 16:167-229.

——. 1977. "The Use of Biological Indicator Organisms to Monitor Trace Metal Pollution in Marine and Estuarine Environments—a Review." *Environ. Pollut.* 13:281-317.

Plewa, M. 1978. "Activation of Chemicals into Mutagens by Green Plants: A Preliminary Discussion." *Environ. Health Perspect.* 27:45-50.

Pollock, E. 1970. "Fertilization in *Fucus.*" *Planta* (Berlin) 92:85-99.

——. 1969. "Interzonal Transplantation of Embryos and Mature Plants of *Fucus.*" In *Proceedings of the 6th International Seaweed Symposium*, ed. R. Margalef, pp. 345-356. Madrid: Subsecretária de la Marina Mercante.

Preston, A., D. Jeffries, J. Dutton, B. Harvey, and A. Steele. 1972. "British Isles Coastal Waters: The Concentrations of Selected Heavy Metals in Sea Water, Suspended Matter and Biological Indicators—a Pilot Survey." *Environ. Pollut.* 3:69-82.

Quatrano, R. 1974. "Developmental Biology: Development in Marine Organisms. In *Experimental Marine Biology*, ed. R. Mariscal, pp. 303-346. New York: Academic Press.

Rao, T. K., J. Young, C. Weeks, T. Slaga, and J. Epler. 1979. "Effect of the Co-Carcinogen Benzo(e)pyrene on Microsome-Mediated Chemical

Mutagenesis in *Salmonella typhimurium.*" *Environ. Mutagenesis* 1: 105-112.

Robinson, J., A. Richardson, A. Crabtree, J. Coulson, and G. Potts. 1967. "Organochlorine Residues in Marine Organisms." *Nature* 214:1307-1311.

Rosin, M., and H. Stich. 1980. "Enhancing and Inhibiting Effects of Propyl Gallate on Carcinogen-Induced Mutagenesis." *J. Environ. Pathol. Toxicol.* 4:159-167.

——. 1979. "Assessment of the Use of the *Salmonella* Mutagenesis Assay to Determine the Influence of Antioxidants on Carcinogen-Induced Mutagenesis." *Int. J. Cancer* 23:722-727.

Seeliger, U., and P. Edwards. 1977. "Correlation Coefficients and Concentration Factors of Copper and Lead in Seawater and Benthic Algae." *Mar. Pollut. Bull.* 8:16-19.

Simmon, V., and R. Tardiff. 1976. "Mutagenic Activity of Drinking Water Concentrate." *Mutat. Res.* 38:389.

Sorenson, F. 1969. "Embryonic Genetic Load in Coastal Douglas Fir, *Pseudotsuga menziesii.*" *Am. Nat.* 103:389-398.

Steele, R. 1978. "Sensitivity of Some Brown Algal Reproductive Stages to Oil Pollution." *Proc. Int. Seaweed Symp.* 9:181-190.

Stich, H., and A. Acton. 1976. "The Possible Use of Fish Tumors in Monitoring for Carcinogens in the Marine Environment." In *Tumors in Aquatic Animals.* Vol. 20 of *Progress in Experimental Tumor Research*, ed. C. Dawe, D. Scarpelli, and S. Wellings, pp. 3-34. Basel: S. Karger.

Stich, H., and U. Kuhnlein. 1979. "Chromosome Breaking Activity of Human Feces and Its Enhancement by Transition Metals." *Int. J. Cancer* 24: 284-287.

Stich, H., L. Wei, and R. Whiting. 1979. "Enhancement of the Chromosome-Damaging Action of Ascorbate by Transition Metals." *Cancer Res.* 39:4145-4151.

Stoltz, D. R., B. Stavric, F. Iverson, R. Bendall, and R. Klassen. 1979. "Suppression of Naphthylamine Mutagenicity by Amaranth." *Mutat. Res.* 60: 391-393.

Sugimura, T., and M. Nagao. 1980. "Modification of Mutagenic Activity." In *Chemical Mutagens. Principles and Methods for Their Detection.* Vol. 6, ed. F. de Serres and A. Hollaender, pp. 41-60. New York: Plenum.

Sunderman, F. W. 1979. "Mechanisms of Metal Carcinogenesis." *Biol. Trace Element Res.* 1:63-86.

——. 1978. "Carcinogenic Effects of Metals." *Federation Proc.* 37:40-46.

——. 1977. "Metal Carcinogenesis." In *Advances in Modern Toxicology.* Vol. 2, ed. R. Goyer and M. Mehlman, pp. 257-295. Washington, D.C.: Hemisphere.

van der Meer, J. 1979. "Genetics of *Gracilaria tikvahiae* (*Rhodophyceae*). VI. Complementation and Linkage Analysis of Pigmentation Mutants." *Can. J. Bot.* 57:64-68.

van der Meer, J., and N. Bird. 1977. "Genetics of *Gracilaria sp.* (*Rhodophyceae, Gigartinales*). I. Mendelian Inheritance of Two Spontaneous Green Variants." *Phycologia* 16:159-161.

van der Meer, J., and E. Todd. 1977. "Genetics of *Gracilaria sp.* (*Rhodophyceae, Gigartinales*). IV. Mitotic Recombination and Its Relationship to Mixed Phases in the Life History." *Can. J. Bot.* 55:2810-2817.

Vig, B. K. 1975. "Soybean (*Glycine max*): A New Test System for Study of Genetic Parameters as Affected by Environmental Mutagens." *Mutat. Res.* 31:49-56.

Yahagi, T., M. Degawa, Y. Seino, T. Matsushima, M. Nagao, T. Sugimura, and Y. Hashimoto. 1975. "Mutagenicity of Azo Dyes and Their Derivatives." *Cancer Letters* 1:91-96.

Yamasaki, E., and B. Ames. 1977. "Concentration of Mutagens from Urine by Adsorption with the Nonpolar Resin XAD-2: Cigarette Smokers Have Mutagenic Urine." *Proc. Natl. Acad. Sci. U.S.A.* 74:3555-3559.

Zimmermann, F., R. Kern, and H. Rasenberger. 1975. "A Yeast Strain for Simultaneous Detection of Induced Mitotic Crossing Over, Mitotic Gene Conversion and Reverse Mutation." *Mutat. Res.* 28:381-388.

6

Detection of Ambient Levels of Mutagenic Atmospheric Pollutants with the Higher Plant *Tradescantia*

L. A. Schairer
R. C. Sautkulis

INTRODUCTION

Nature and Extent of the Pollution Problem

As a result of very great increases in world population and expansion of industrial and agricultural development, human beings are exposed to an ever increasing level of environmental pollutants and/or mutagens. The sources of these chemical and physicial contaminants of the environment are varied. They include human and industrial pollution of streams and groundwater (Brittin, 1973; Klekowski, 1978), automotive and industrial combustion products (Bond and Straub, 1973), pesticides (White-Stevens, 1971; Epstein and Legator 1973), commercially used chemical additives, solvents, or catalysts (Fishbein 1973), pesticides (White-Stevens, 1971; Epstein and Legator, 1971), sources of radiation (Eisenbud, 1973; Mericle and Mericle, 1965; NCRP, 1980). In the past few years there has been an accelerating effort to detect and identify the most toxic pollutants and to assess their real or potential hazard to biological systems (Flamm, 1974; Flamm and Mehlman, 1978; Hindawi, 1970; Oser, 1971; Vogel and Röhrborn, 1970; Neel, 1974; Hughes et al., 1980). It is to this end that the very sensitive *Tradescantia* plant has been exploited as a somatic mutation test system to study the genetic effects of physical and known or suspected chemical mutagens.

Historical Perspective of the Development of the *Tradescantia* Stamen Hair Bioassay

Extensive use of *Tradescantia* hybrids in radiobiological studies has provided much information about their radiosensitivity with respect to somatic mutation produced by X, gamma, beta, neutron, muon, and heavy ion radiations (see Table 6.1). Data from combined radiation and weightlessness have been obtained by scientists from both the USSR and the United States in experiments conducted in space vehicles (Delone et al., 1968; Sparrow, Schairer, and Marimuthu, 1968), and many data exist on induced chromosome aberrations (see review by Savage, 1975). Our studies have included investigations of correlations between chromosomal aberrations and

TABLE 6.1
Typical Mutagenic Responses of *Tradescantia* Clone 02 Following Exposure to Various Radiations

Radiation	*Dose(rads)*	*Hairs Scored* ($\times 10^3$)	*Pink Mutations/1,000 Hairs* - Control** ± S.E.*	*References*
Gamma, cesium-137 (chronic 18/20°C)	2.8/d	47	15.60 ± 0.72	Sparrow, Schairer, and Marimuthu (1971)
(chronic 13°C)	2.8/d	8	24.4 ± 2.53	Sparrow, Schairer, and Marimuthu (1971)
(chronic 13°C)	0.72/d	17	4.7 ± 0.98	Sparrow, Schairer, and Marimuthu (1971)
$Beta_1$ ^{90}Sr–^{90}Yr (acute)	1.0	7	1.0 ± 0.38	Bottino, Bores, and Sparrow (1973)
X rays, 250-kVp (acute)	0.25	961	0.17 ± 0.06	Sparrow, Underbrink, and Rossi (1972)
Neutrons, 0.43-MeV (acute)	0.01	96	0.44 ± 0.17	Sparrow, Underbrink, and Rossi (1972)
Nitrogen ions, 3.9-GeV(acute)	0.3	29	2.53 ± 0.63	Underbrink, Schairer, and Sparrow (1973a)
Muons, $^+$6-GeV (acute)	12.03	19	2.56 ± 0.39	McNulty, Sparrow, and Schairer (1974)

*Induced mutation rates are significantly above control rates at 1 percent level.
**Cumulative control value per 1,000 hairs for clone 02 is 0.65 ± 0.07.

somatic mutations (Sparrow, Schairer, and Marimuthu, 1971). More recent work has demonstrated that the methods used to study the genetic or cytogenetic effects of physical mutagens can be applied directly to chemical mutagen studies (Underbrink, Schairer, and Sparrow, 1973b; Nauman, Sparrow, and Schairer, 1976). Data from exposures to ionizing radiation can be used as a basis of quantitative comparison of the effects of chemical mutagens (Sparrow, Schairer, and Villalobos-Pietrini, 1974), and as models for the slopes and shapes of dose-response curves associated with exposure to chemical mutagens.

The methodology for radiobiological studies has evolved over a period of more than 50 years, commencing with the pioneer irradiation work with *Drosophila* by Muller (1927) and with maize by Stadler (1928). The development of the atomic bomb and related nuclear detonations, followed by the search for peaceful uses of radiation in the 1950s and 1960s, focused attention on the environmental impact of radiation (Sparrow, 1960; Singleton, 1958; Fowler, 1965). New, more sophisticated methods of radiation detection and measurement were developed (Attix, 1969; 1972; Knoll, 1979). These new dosimetric techniques permit radiation dosage determinations for exposures to the whole body or to single molecular targets for a wide range of radiation types and energies. Environmental effects of radiation from airborne particulates were assessed through plume studies of atmospheric dispersion of particulates (Norman and Winchill, 1971), food chain contamination (Russell, Bartlett, and Bruce, 1971), and impact on human health (Larsen et al., 1978).

In the late 1960s it became apparent that the energy and industrial chemical revolution in the world was generating a whole new spectrum of pollutants in the environment, both in the workplace and at home. Epidemiological studies began to show a correlation between industrial effluents and certain human health problems (Epstein, 1974; Higginson, 1968; Heath, 1978; Creech and Johnson, 1974); marine and terrestrial fauna and flora were showing toxic effects of exposure to polluted atmospheres and streams (Commoner, 1971; Skelley, 1980). Hence, the biological responses to these chemical environmental stresses are now documented and no longer a matter of speculation. The ambient level of genotoxic agents in the atmosphere has generated an environmental impact that is real and not hypothesized, as was the circumstance when the early use of radiation was being challenged. The effect of chemical environmental mutagens on vegetation, food chain, and human health is a real problem, and one requiring a whole new technology for both quantitative measurement and biological assessment.

Many academic institutions, national laboratories, and private companies have initiated research programs to develop systems to assess the potential biological hazard of existing environmental pollution or to demonstrate the absence of genotoxic effect of new compounds (depending upon discipline or mandate). In 1969 the Environmental Mutagen Society was founded to encourage the study of mutagens in the human environment, to engage in and to sponsor research, and to disseminate information related to this problem. Research with a wide range of organisms including prokaryotic microbes, eukaryotic higher plants, eukaryotic mammalian cells in vitro, and whole animals ensued from the early 1970s to the present. The state of the art of many of these bioassays has been reviewed by de Serres and Shelby (1978) and Hollaender and de Serres (1971-1980).

The bioassays described in these reviews, and currently available, are primarily directed toward assessment of environmental chemical pollutants in the liquid or particulate phase. With the exception of the higher plant *Tradescantia*, few, if any, have demonstrated the potential for assessing the genotoxic effects of vapor-phase pollutants, and certainly not in situ at ambient levels. The stamen hair somatic mutation assay has been used successfully to detect mutagenic levels of pollutants in ambient atmospheres near several industrial complexes without the necessity of collection, concentration, and elution of compounds. The primary objective of this chapter is to present the state of the art of the *Tradescantia* stamen hair system, including a physical description of the plant, the genetic basis for the mutant character used, radiobiological background, validation of the system (using laboratory exposure to aqueous and gaseous known mutagens), and the application of the stamen hair assay to ambient air monitoring.

THE *TRADESCANTIA* STAMEN HAIR BIOASSAY

Background for the Selection of the Genus *Tradescantia*

Much of the radiobiological research was pioneered by Dr. Arnold H. Sparrow in the early to mid 1950s (Sparrow, 1951; Sparrow and Singleton, 1953; Sparrow and Gunckel, 1956; Sparrow, 1962; Sparrow, Schairer, and Sparrow, 1963). Sparrow extended his studies on factors influencing the radiosensitivity of plants from lethality to somatic mutation (Sparrow and Pond, 1956; Cuany,

Sparrow, and Pond, 1958). Localized phenotypic changes in petal color following X-irradiation and acute or chronic gamma-irradiation have been summarized by Sparrow et al. (1968), and data were obtained showing that the *Tradescantia* flower color locus is far more sensitive to radiation than other genera tested—*Tulipa*, *Clematis*, *Trapaeolum*, *Gladiolus* (Figure 6.1). Besides the advantage of the high radiosensitivity of *Tradescantia*, the genus was also better suited to cytological analysis than other genera because of its low number (2n = 12) of relatively large chromosomes (see Figure 6.2). These two major advantages, coupled with its ease of vegetative propagation; vigorous growth habit under greenhouse, growth chamber, and field conditions; and year-round flower production led

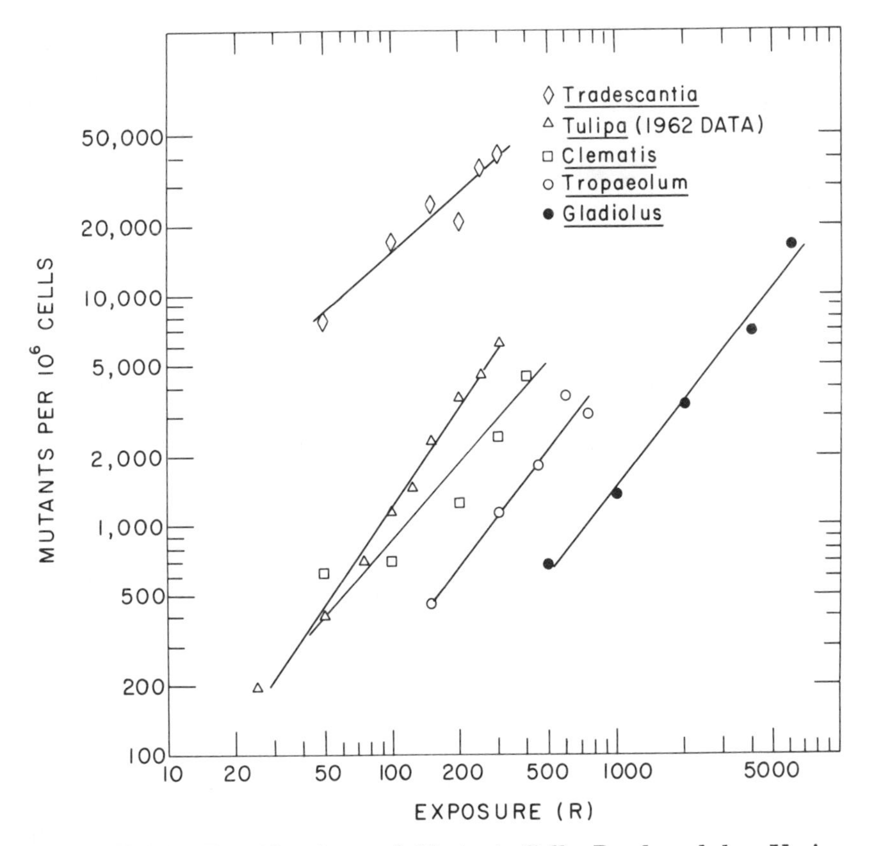

FIGURE 6.1. The Number of Mutant Cells Produced by Various Exposures of Acute Gamma Irradiation over a 16-Hour Period (Control Rates Subtracted).

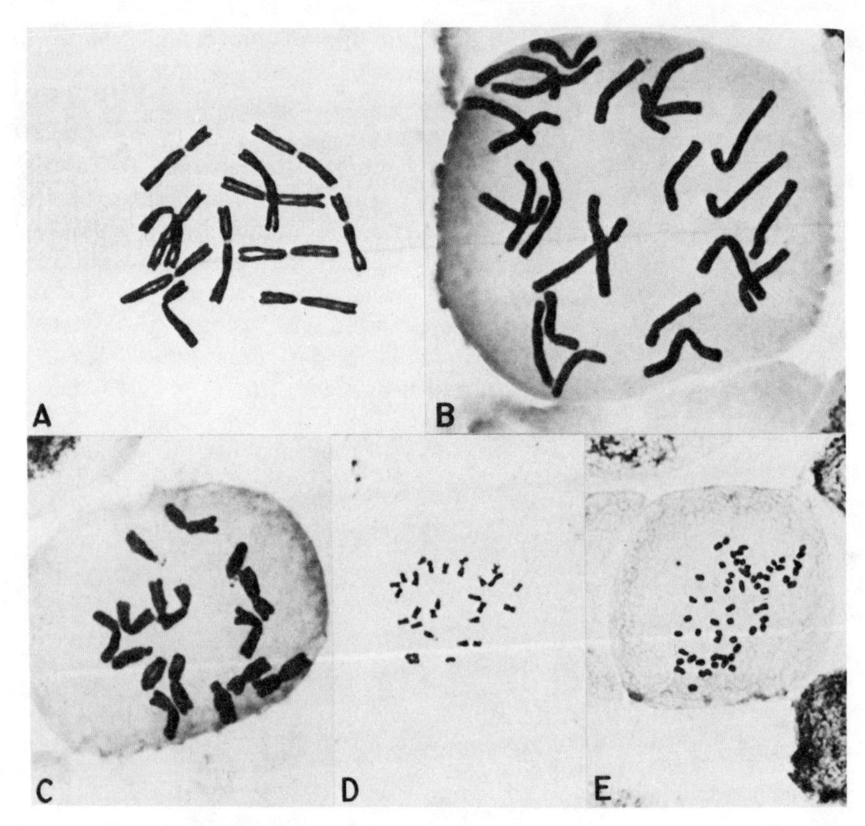

FIGURE 6.2. Root Tip Metaphase Chromosomes of (A) *Tradescantia* clone 02, (B) *Tulipa* HV Golden Harvest, (C) *Clematis jackmanii*, (D) *Tropaeolum majus*, and (E) *Gladiolus* HV Mansoer. All magnifications 648 X. Source: Sparrow et al. (1968).

to its extensive use and further development, initially as a radiobiological tool and subsequently as a chemical mutagen assay.

Taxonomic Background and Genetic Basis for the Mutant Phenotype

The early radiobiological studies cited above were made using *Tradescantia* hybrid clone 02, shown in Figure 6.3. Clone 02 was discovered in 1958 by Dr. A. H. Sparrow as a seedling in an outdoor collection of *Tradescantia* assembled by Professor W. J. Brown at the University of Texas at Austin. The parental species are unknown,

FIGURE 6.3. Normal Stock Plants of *Tradescantia* Clones 02 and 4430 Showing Several Mature Inflorescences. Insert: Close-up of a mature inflorescence showing a full range of buds.

but are thought to be *T. occidentalis* and *T. ohiensis* (Mericle and Mericle, 1967). The combined factors of unknown parentage, self-fertility leading to potential seedling contamination of stock population, high and variable spontaneous rate of pollen abortion (30-50 percent), and uncertainty of single-locus genetic control of the phenotypic change in flower color (due to failure of self-fertilization to yield expected 3:1 segregation) engendered uncertainty as to the desirability of continued use of this clone. These disadvantages prompted the development of the *Tradescantia* clone designated 4430, a hybrid between blue-flowered *T. hirsutiflora* (2461C) and pink-flowered *T. subacaulis* (2441), which is shown in Figure 6.3 (right) and was described in detail by Thompson-Emmerling and Nawrocky (1980). The genetic basis for the observed pink-celled mutations was confirmed by test crosses of clone 4430 with its pink-flowered parent and crosses between clone 4430 and two heterozygous sibling F_1 plants; these crosses resulted in classes 1:1 and 3:1 segregation ratios. The results of these crosses are consistent with the hypothesis that pink pigment is determined by a pair of alleles at a single locus, and that blue (B) is dominant to pink (b) (Thompson-Emmerling and Nawrocky, 1980).

In addition to the genetic documentation, other advantages of the clone 4430 include more stable meiosis, low spontaneous pollen abortion rate (10 percent), self-sterility (facilitating seedling-free stock populations), good flower color contrast in mutant cells, and greater sensitivity to chemical mutagens than clone 02 (Thompson-Emmerling and Nawrocky, 1980). On the basis of these advantages, *Tradescantia* clone 4430 has been used exclusively in all the ambient air studies initiated at the Brookhaven National Laboratory.

Description of the Test System

The inflorescence is determinate and composed of 18-20 flower buds arranged as a gradient of developmental age, with the oldest topmost on the inflorescence (Figure 6.3, insert). Each stamen hair is a filament derived from a single epidermal cell that grows primarily through successive divisions of the apical and subapical cells (Figure 6.4[a]). At maturity and under optimal growth conditions, untreated stamens have 40-75 hairs, each with an average of 24 cells (Figure 6.4[b]). A mutant cell produced early in development is capable of having several mutant progeny that appear in the flower as a string of pink cells (Figure 6.4[c]). Because of this cell multiplicity factor,

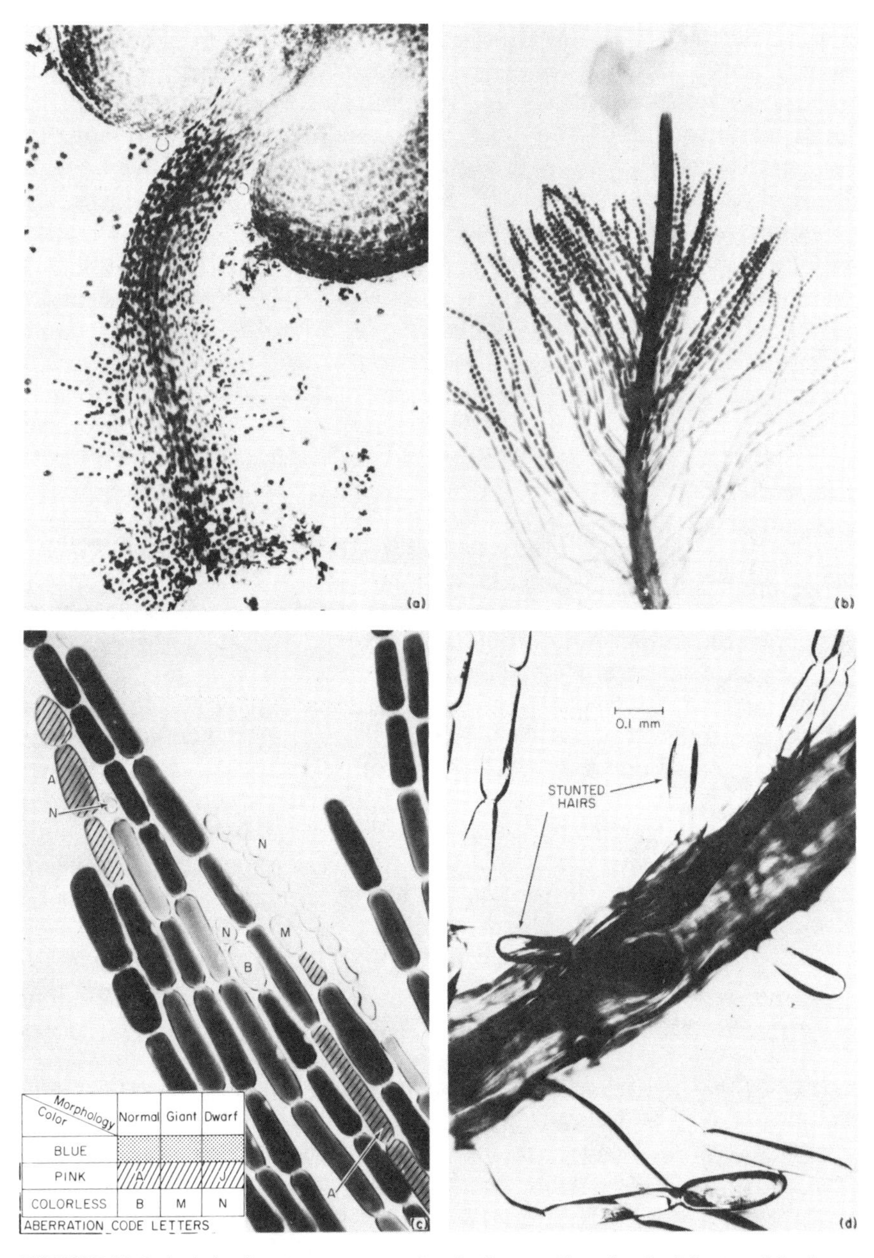

FIGURE 6.4. (a) A stamen excised from the bud 13 to 15 days before anthesis, with very young developing stamen hairs. (b) Mature stamen showing stamen hairs spread out on a slide ready for scoring for mutated cells. (c, d) Mutant or aberrant cells or hairs typical of those seen after irradiation.

scoring each cell in the string as a mutant would be erroneous; hence, a contiguous string of pink cells is noted as a single event.

The flower is trimerous, so that counting hairs on one of each of the three antipetalous and antisepalous stamens provides an index of hair number for an average flower. When exposed to chemical or physical mutagens, some meristematic cells are mutated. On the other hand, if the lesion produced is lethal, cell division would cease and a stunted mature hair would result (Figure 6.4[d]). By counting the number of cells in the shortened hair, a survival curve can be constructed that is useful to the interpretation of mutation dose-response curves.

Plant Culture and Posttreatment Data Collection and Analysis

Mature plants vary from 20 to 45 cm in height. They produce many flowering shoots or branches, each with an inflorescence composed of a series of flower buds covering a range of ages such that once blooming starts, roughly one flower blooms per day.

For experimental use, large plant populations are grown in controlled environment chambers under artificial light from fluorescent and incandescent lamps. An 18-hour day is maintained with a light intensity of about 1.8×10^4 lux from Sylvania FR96 T12/CW/VHO/135° bulbs supplemented by about 20 percent incandescent light under day/night temperatures of 20° ± 0.5 and 18° ± 0.5C. When exposures to a mutagen are scheduled, fresh cuttings are made, each bearing a young inflorescence, and treated within a few hours. If prolonged observations are required, potted plants can be used also. After exposure the cuttings are placed in containers of aerated Hoagland's nutrient solution and kept in a controlled-environment growth chamber under the standard conditions described above until observations are completed (Figure 6.5).

The posttreatment scoring period is important because the assay system is a determinate developing system. A daily plot of the frequency of induced pink events found during the several weeks following an acute exposure to a mutagen has three stages: an initial lag period of low, approximately background frequency; several days of increasing frequency, reaching a peak or plateau; and a subsequent declining frequency (Figure 6.6).

The reasons for the three stages of mutant frequency, discussed by Kudirka and van't Hof (1980), appear to be associated with the fact that bud arrangement in the inflorescence is characterized by a

FIGURE 6.5. ***Tradescantia*** **Clone 4430 Floral Cuttings Prepared for Exposure, 60 Cuttings per Dish.**

size gradient that reflects developmental age. In the younger buds the larger portion of the cells are in G_1, and the others are distributed in S, G_2, and M. There is an inverse relation between cell number and the proportion of cells in mitosis with increasing bud age. Likewise, the number of cells in S first decreases with age but then, at an intermediate age when mitosis ceases, it rises again just prior to pigmentation. This last wave of DNA synthesis signals the mass movement of the cells from G_1 to G_2, where they arrest. Hence, during bud development there are two periods of DNA synthesis: an early one in younger buds, associated with mitosis, and a later one that occurs as cell division ceases, 5-12 days before blooming. Both periods of DNA synthesis occur prior to pigmentation, which begins about three days before blooming (Figure 6.7).

Given these cellular changes and an acute treatment with a chemical mutagen that is most effective during DNA synthesis, the induced mutant frequency curve should have three stages: first,

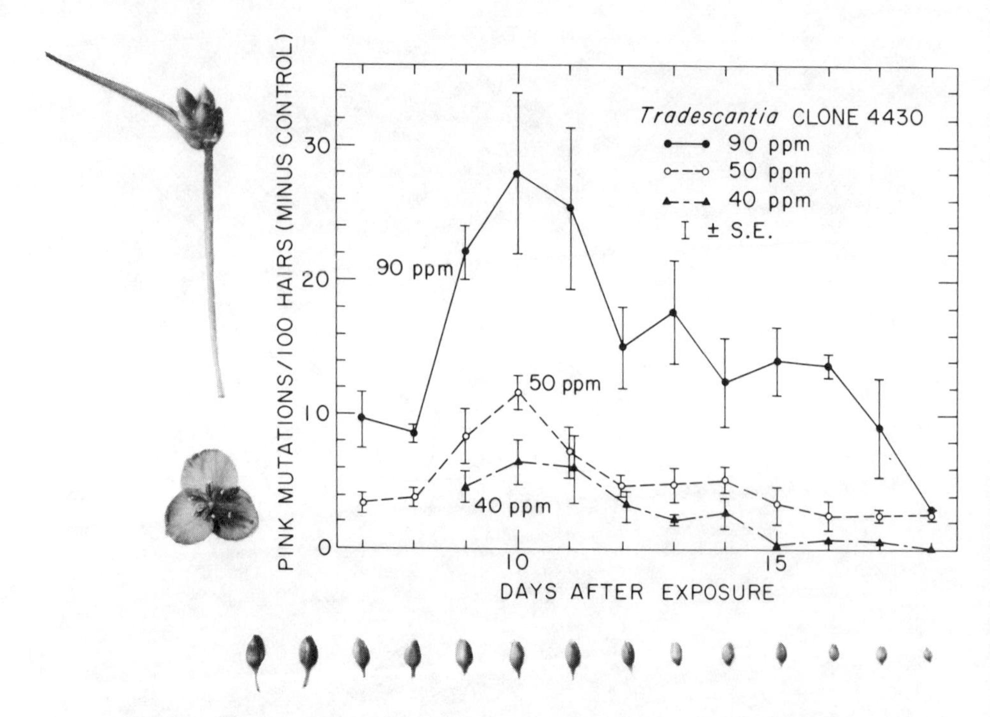

FIGURE 6.6 Pink Mutation Rates in *Tradescantia* Stamen Hairs Following 6-Hour Exposure to Gaseous EMS Reach Peak Values 7–12 Days after Exposure. Note: Pink events are analyzed in mature flowers; the corresponding bud sizes at time of treatment are shown along the abscissa. Source: Schairer et al. (1978), with permission.

in the older buds a low frequency that reflects the fact that they were pigmented before treatment; second, a peak frequency composed of cells that were progressing from G_1 and G_2 in buds of an intermediate age; and third, a decline to a plateau that represents previously proliferating cells that were in the S period when the buds were very young. Thus, the shape of the mutant frequency curves after a chemical treatment mirrors the DNA synthetic activity of the target cells (Figures 6.6 and 6.7).

The first rule to follow in dosage selection is to avoid killing the cells. Lethal doses are determined by counting the number of cells per hair. If there are fewer than 23, the hair is stunted and the dose is too high (Figure 6.4[d]). Selection of suitable doses nearly always is empirical, but Sparrow, Schairer, and Villalobos-Pietrini (1974) settled on a standard treatment time of six hours at concentrations

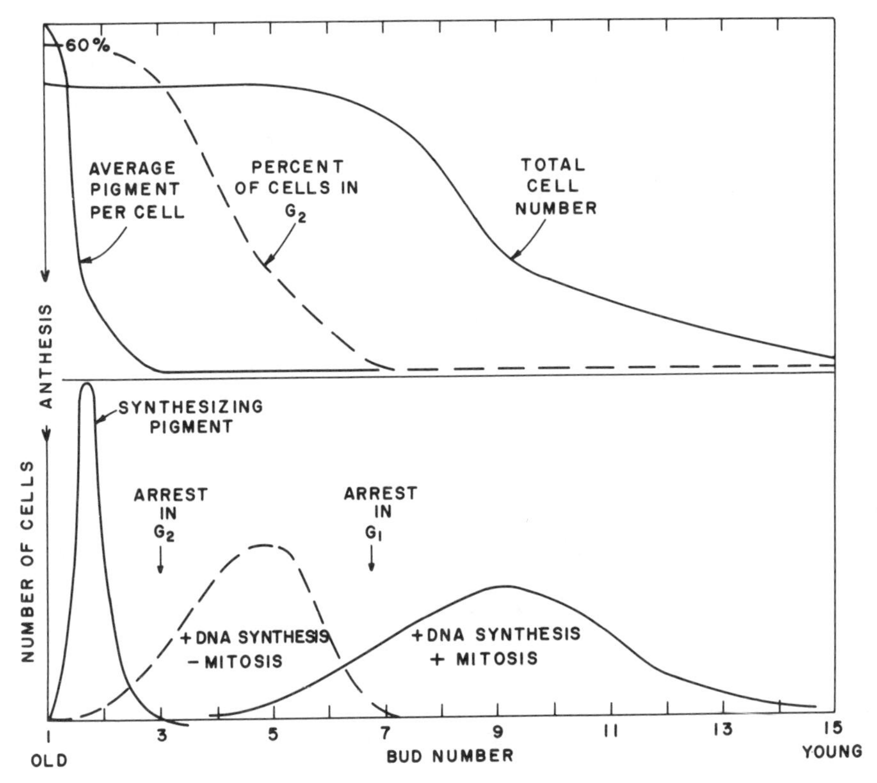

FIGURE 6.7 Summary of the Cell Kinetics for Developing Flower Buds in *Tradescantia* Clone 4430. The relationship between bud age, cell number, G_1, DNA synthesis, G_2, and pigment formation is shown. Source: Kudirka and van't Hof (1980).

between 3 and 150 ppm of mutagen for potent compounds such as EMS and 1,2-dibromoethane.

The mutation data are collected over a period of five to seven days. The extended posttreatment time is mandated by the developmental biology of the inflorescence, as described above. During the five to seven days, the flowers are collected daily as they bloom; then the stamens are removed, mounted for microscopic examination, and scored for pink cells among the blue ones. The number of mutant events per flower, number of hairs, treatment, date, and other pertinent details are recorded directly on computer input sheets for analysis. The data gathered are expressed as the number of pink mutant sectors per 100 or 1,000 hairs. The mean number of mutant events per hair is calculated by summing the number of pink events from individual flowers blooming during the five- to seven-day peak response period (Figure 6.6) and dividing by the product of total flowers analyzed times the estimated number of hairs per flower. All data are processed via a computer program written by K. H. Thompson for the CDC 6600 and 7600 computers at Brookhaven National Laboratory.

To present data as a dose-response curve after acute and chronic exposures, a mean peak value is used on a log-log plot of mutant frequency versus radiation dosage in rad or chemical concentration in ppm. In every experiment both treated and untreated plants are scored, so the tabulation of mutant frequencies includes spontaneous and induced mutations. The net mutant frequency of the treated cells is then obtained by subtraction of the background (untreated) frequency. For chamber-grown clone 4430 the background mutation frequency is 3.35 ± 0.09 per 1,000 hairs.

REVIEW OF THE RADIOBIOLOGY OF THE SOMATIC MUTATION SYSTEM

Acute Exposures to Radiations of Various Qualities

Many kinds of ionizing radiations have been used including X rays, gamma rays, monoenergetic neutrons of various energies, muons, pions, and heavy particles such as nitrogen ions. Table 6.1 summarizes some mutation rates for low doses obtained with several of these physical mutagens, and Figure 6.8 demonstrates graphically the relative biological effectiveness of these various radiations.

Typical dose-response curves for clones 02 and 4430 for pink mutations per stamen hair, plotted against X-ray dose in rads, are shown in Figure 6.9. These dose-response curves for 250-kVp X rays are used as standards of comparison for all of our radiation experiments and for comparison with mutation rates induced by various chemical mutagens. It can be seen that the two curves demonstrate the good linearity and extreme range in response over at least three log cycles. The high sensitivity of the stamen hair system is shown by the significant response below 0.25 rad with no indication of a threshold dose. Note that at low doses there is a +1 slope, which becomes somewhat steeper above about 6 rads, and that the curves bend over and then decrease at high doses. These curves will be shown in succeeding figures for comparison with chemical mutagen responses.

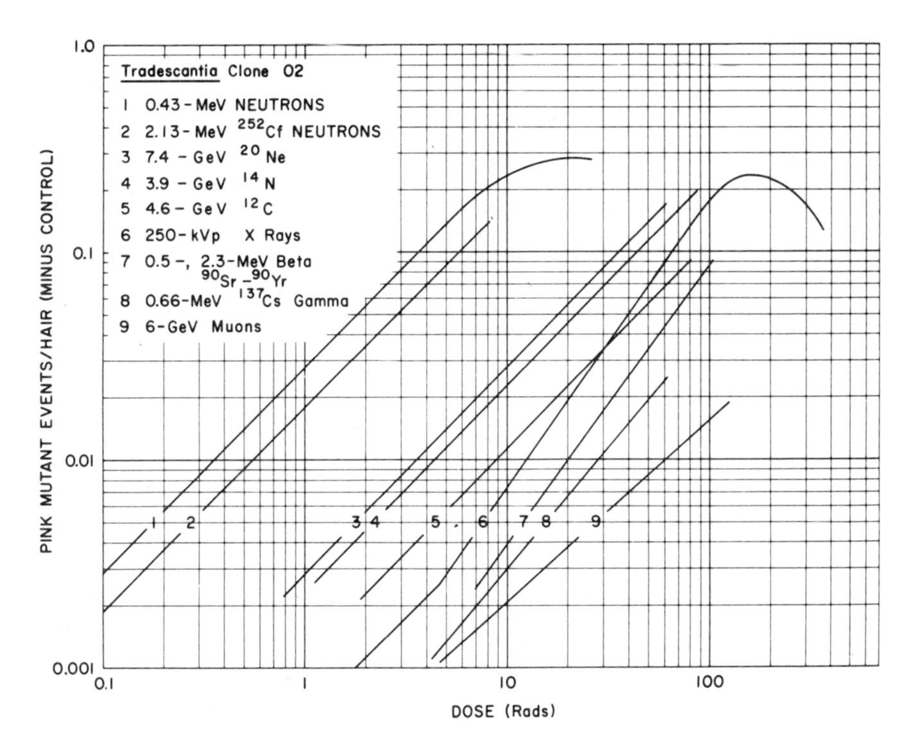

FIGURE 6.8 Dose-Response Curves of *Tradescantia* Clone 02, Representing the Relative Biological Effectiveness of the Various Radiations Indicated.

Chronic Exposure to Radiation

An even more important response is that resulting from chronic irradiation, since most exposures to air pollutants are chronic. Data from chronic gamma exposures of clone 02 indicate an accumulation of pink events for the first three weeks, followed by a plateau for as long as the radiation continues (Figure 6.10). These data suggest that the maximum sensitivity for the stamen hair system is obtained after about three weeks of exposure and that the plateau response values are proportional to exposure rates and not to total dose.

The high sensitivity of the flower-color locus and the physical hardiness of the plant were demonstrated further by the chronic in situ exposures of clone 02 to natural background radiation (0.25mR/hr) from a radioactive dike in Colorado. The roots of the test plants

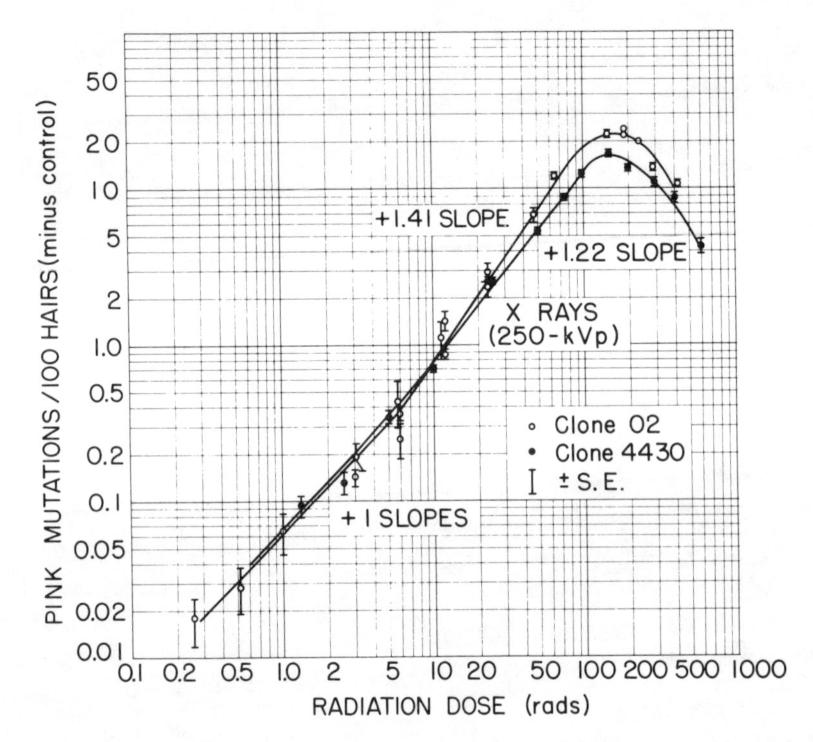

FIGURE 6.9. Log-log Plot Showing Great Similarity in the Frequency of Pink Events in Stamen Hairs of *Tradescantia* Clones 02 and 4430 Following Acute X-ray Exposures. Source: Sparrow et al. (1974), with permission.

were isolated from field soil so that the source of exposure to the inflorescences was external to the plant. The results of these studies by Mericle and Mericle (1965) were a four- to fivefold increase in somatic mutation frequency from field studies and a 1.6-fold increase from comparable laboratory studies. These data support the use of the *Tradescantia* test system as an in situ monitor for external background radiation.

In contrast, the efficacy of internal exposure from isotopes taken up through the roots was demonstrated by data from a pilot study completed at the Brookhaven National Laboratory. In this experiment there were daily applications of tritiated water, at concentrations of 3 and 12 microcuries per ml, respectively, to the soil of two sets of potted plants of clone 4430 in a growth chamber. Daily flower analysis showed a maximum increase in somatic mutation frequency

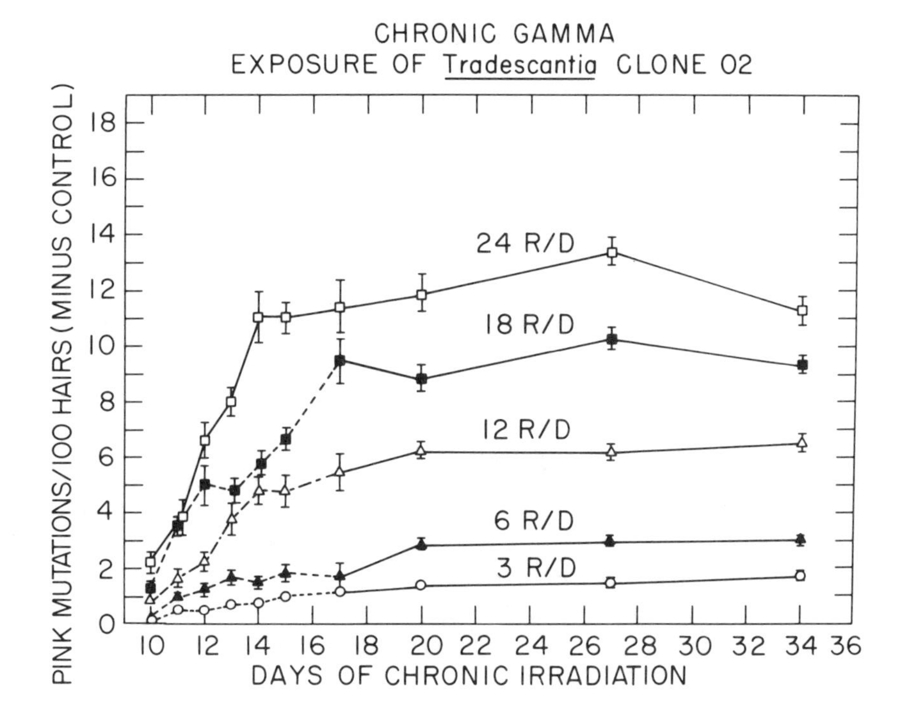

FIGURE 6.10. The Frequency of Pink Events Produced in the Hairs of Clone 02 during Chronic Gamma Exposure, Plotted against Days of Chronic Irradiation. Source: Schairer et al. (1978), with permission.

after about three weeks of treatment. Calculated beta radiation dose rates of 0.87 and 3.48 rads per day induced stamen hair mutation rates of 11 and 15 per 1,000 hairs, respectively (see Table 6.2). These data support the suitability of the stamen hair system for use as a monitor of both external and internal radiation, as well as demonstrating the capability of mutagen uptake through the roots.

TABLE 6.2
Mutagenic Response of *Tradescantia* Clone 4430 to Chronic Internal Beta Radiation from ^3H_2O

Mutagen	*Dose Rate (rad/day)*	*Flowers Analyzed*	*Hairs Analyzed*	*Mutant Events*	*Mutant Events/ Hair - Control ± S.E.*
^3H_2O	0.87	625	252,731	2,744	0.01087 ± 0.00021
^3H_2O	3.48	10	3,189	48	0.01505 ± 0.00237

CHEMICAL MUTAGEN EXPOSURES UNDER LABORATORY CONDITIONS

Development of Exposure Techniques for Nonvolatile Compounds

Tradescantia cuttings can be exposed to liquid mutagens by immersion of the inflorescences or to gaseous mutagens by fumigation in open air or in gastight chambers. Exposure to gaseous chemical mutagens is a relatively new phase of mutation research, requiring the development of new exposure and dosimetric techniques.

Nonvolatile aqueous chemicals can be tested most effectively by simply immersing the entire inflorescence in solutions of desired concentrations. If the chemical to be tested is a solid, the most suitable solvent is water or 1 percent dimethylsulfoxide (DMSO). The exposure may be for a few minutes or as long as 24 hours without interfering with the normal posttreatment development of the inflorescence. Exposure time is usually determined by the potency of the mutagen or its overall toxicity. During the exposure the base of the cutting must be kept as wet as possible; water-soaked absorbent cotton may be used. Another approach is to use potted plants with floral stalks long enough to allow bending and direct immersion of the inflorescence. Surface residues are removed by rinsing the inflorescence with water or an appropriate detoxifying agent. All treatment solutions and residues must be handled according to accepted safety procedures (Calbiochem-Behring Corp.).

The use of nonvolatile chemical mutagens has played an essential role in the validation of the *Tradescantia* stamen hair test system. Many of the more common chemical mutagens have been used in solution because of their low volatility and, more important, because nearly all other bioassays require a liquid phase for infusion into sterile culture media, direct injection into the organism, or imbibition (such as seed soaking). Therefore, in order to validate the stamen hair system as a mutagen and/or carcinogen detector, the well-documented liquid mutagens must be used. Some typical results of inflorescence immersion are given in Table 6.3.

Development of Exposure Techniques for Gaseous or Airborne Chemicals

Monitoring gaseous chemicals for mutagenicity required a concept different from those developed for use with ionizing radiation.

TABLE 6.3
List of Chemicals to Which *Tradescantia* Clone 4430 Was Exposed by Inflorescence Immersion in Liquid at the Concentrations Indicated

Chemical	*Exposure Time (Hrs.)*	*Minimum* Conc.*	*Maximum** Conc.*	*Statistical Signif. (Percent)*
Caffeine	chronic		10^{-3}M to stem	NS
Atrazine	chronic		0.045 gm/pot	NS
Atrazine	24	6.90×10^{-5}M		1
Sodium azide	3	10^{-4}M		1
Dimethylamine hydrochloride	2	10^{-2}M		10
N-methyl-N-nitro-N-nitrosoguanidine	1.5	10^{-3}M		1
Safrole	24	6.17×10^{-6}M		1
$3,3^1,5,5^1$-tetramethyl benzidine	24	4.16×10^{-6}M		1
Alpha-naphthylamine	24	6.75×10^{-5}M		1
Beta-naphthylamine	24	6.98×10^{-5}M		1
Benzidine	24	5.43×10^{-7}M		1
Hexamethyl-phosphoramide	24	5.58×10^{-5}M		1
Pyrene	24	4.94×10^{-6}M		1
Diethylstil-bestrol	24	3.73×10^{-4}M		1
Benzo(α)pyrene	24	3.96×10^{-5}M		1
N,N^1-ethylene-thiourea	24	9.79×10^{-5}M		1
Methylene chloride	24	1.17×10^{-3}M		1
Diallate	24	9.05×10^{-5}M		1
Triallate	24	9.05×10^{-5}M		1
Methyl methane-sulfonate	24	9.09×10^{-5}M		1
2-nitrofluorene	24	4.74×10^{-6}M		2

Note: The mutagenic effect is given in terms of the level of statistical significance (percent).

*Lowest concentration tested that showed a mutagenic response.

**Highest concentration tested at which no effect was observed.

Exposure to airborne pollutants in laboratory chambers, the workplace, and the outside environment required an entirely new set of dosimetric procedures. The source and identity of the genotoxic agent are commonly unknown (certainly in the ambient atmosphere); gas chromatographic and wet chemistry procedures are required for qualitative and quantitative chemical determinations; the dosage to the target cells is influenced by the biological half-life in air and in wet tissue; translocation of the chemical from surface exposure to target cell is influenced by temperature, day/night light cycle, molecular size, pH, and other factors.

Two approaches to gaseous chemical exposures (static and flow-through) have been made at Brookhaven National Laboratory. The first is a static exposure in which a measured amount of chemical is sealed in a chamber with the cuttings and allowed to evaporate. The air in the chamber is continuously mixed with a fan, and the chemical concentration is assumed to be a direct calculation based on a known amount of chemical in a known volume of air. One potential error in this calculation is that no estimate can be made of the amount of chemical adhering to the walls of the chamber and plant surfaces, or absorbed by the water or culture medium. One advantage of the static exposure technique is the ease of incorporation of vacuum injection to enhance uptake of the chemical. (The procedure will be described below.)

The second approach to gaseous chemical exposures is called the flow-through or dynamic method. When attempting atmospheric mutagen monitoring, the primary characteristics are the long period of exposure (chronic) in the workplace or at home and the infinite pollution source of the ambient atmosphere. To best simulate chronic exposures experienced in our environment, a glass chamber (nine-liter desiccator) is used to contain up to 60 cuttings during exposure (Figure 6.11, upper photo). Filtered air is bubbled through the liquid mutagen in the first impinger tube, saturating the air; excess mutagen is deposited in the second impinger tube. The maximum chemical concentration is determined by the vapor pressure of the compound and is controlled by varying the air flow rate through the liquid mutagen and by dilution of the mutagen-saturated atmosphere with filtered "clean air." A steady flow of gaseous mutagen is maintained through the chamber, with the concentration measured at the intake port and from within the chamber through solenoid-activated sampling valves connected to a gas chromatograph. The gas is exhausted through a copper oxide furnace at 675°C (Kusnetz, Saltzman, and Lanier, 1960), shown at the extreme right in Figure 6.11 (upper), which oxidizes the mutagen, rendering it harmless. The oxidized effluent is then exhausted to the outdoor atmosphere through a 90-meter stack. Thus, the plants may be exposed concurrently in separate chambers to several constant levels of gaseous chemical mutagen for a period of a few hours, several days, or even weeks (Figure 6.11, lower). This fumigation technique simulates

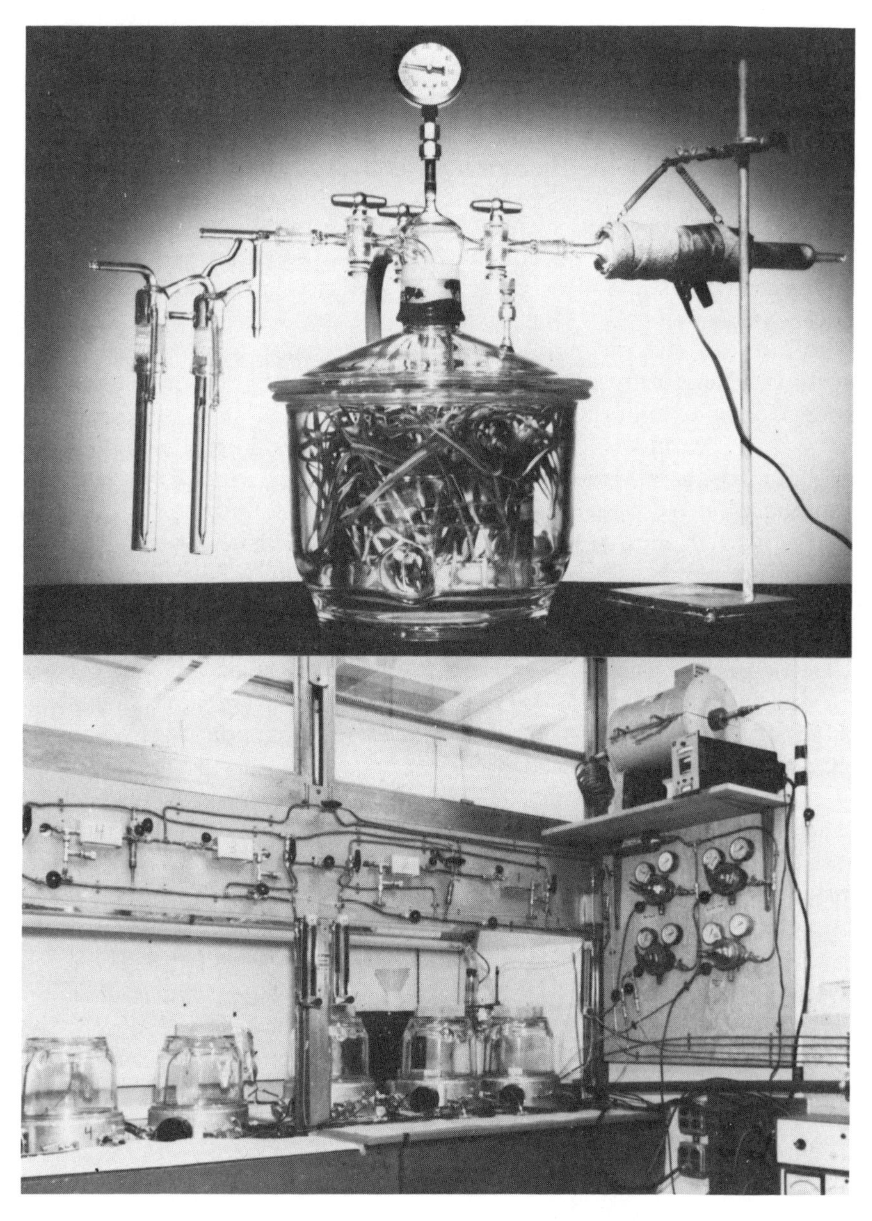

FIGURE 6.11 (Upper) Early Exposure Chamber, a Modified Nine-Liter Desiccator Used for Treating *Tradescantia* Cuttings with Gaseous Chemical Mutagens. (Lower) Multiple Exposure Chamber System Facilitating Simultaneous Gaseous Chemical Treatments at Four Different Concentrations and One Control.

the infinite pollution source of chronic exposure to the ambient atmosphere, and minimizes the problems of chemical sedimentation, absorption, adsorption, biological half-life, and so on.

Factors Influencing the Uptake of Chemical Mutagens

Dosimetry performed with the gas chromatograph provides a measure of the chemical concentration at the surface of the plant. However, it was not clear what true dose would be received by the target tissues (that is, the stamen hair cells in which pink mutations were induced and ultimately registered). A study was made to investigate the kinetics of chemical penetration through the inflorescence tissues, using tritiated 1,2-dibromoethane (DBE) as an indicator of tissue dose (Nauman, Klotz, and Schairer, 1979). In this study *Tradescantia* cuttings were exposed to vapors of [^{3}H]-DBE in the manner described above for gaseous exposures, and flower buds were collected for analysis at various times after treatment. Stamens, petals, sepals, and other parts were excised from flower buds 30 minutes after exposure and were analyzed for tissue content of [^{3}H]-DBE. The data presented in Figure 6.12 indicate that mutagen penetration is directly proportional to time and concentration of exposure, even for exposures as short as 15 minutes. Hence, the stamen hair system is suitable for monitoring airborne chemical pollutants even if exposures are of short duration.

The rate or efficiency of penetrance of DBE is dependent to some degree on the aperature of stomatal cells in epidermal leaf and sepal tissues. Mutation response was more than doubled when inflorescences were exposed to DBE during the light cycle (when stomata were fully open), compared with the rate following exposure during the night cycle (stomata closed). Representative data are shown in Table 6.4.

As mentioned above, chemical uptake can be enhanced by the incorporation of vacuum injection techniques. When using a chemical mutagen such as ethyl methanesulfonate (EMS), more complete tissue permeation is achieved by evacuating the air from the exposure chamber to about 30 mm Hg for approximately 10 minutes, then slowly returning it to 1 atmosphere of pressure by bubbling filtered air through an aqueous reservoir of the chemical mutagen (see Figure 6.11, upper). The liquid chemical mutagen is in the glass impinger tube located at the left of the exposure chamber. Data presented in Table 6.5 show that vacuum injection can increase chemical uptake and, hence, mutation response by about 50 percent.

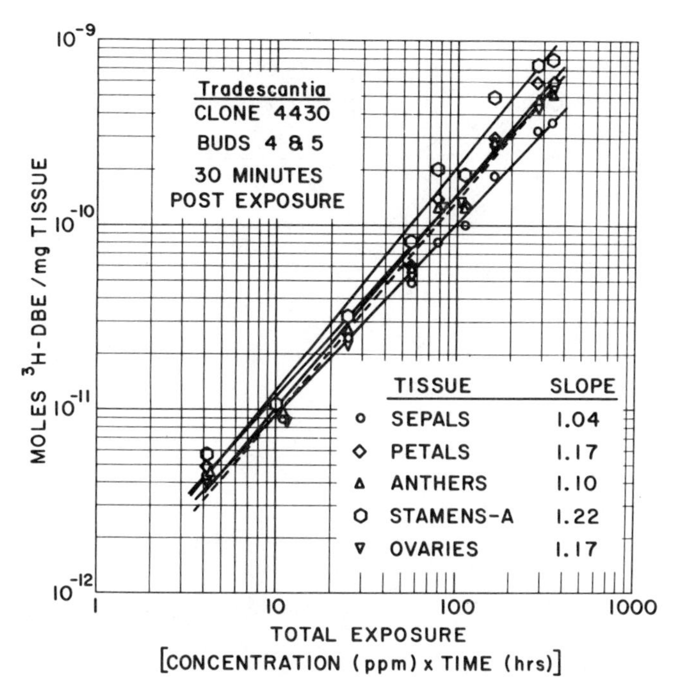

FIGURE 6.12. Exposure Response Curves of Five Flower Parts of *Tradescantia* Showing the Relationship of the Concentration of ^{3}H-DBE to Total Exposure.

TABLE 6.4
Effect of Stomatal-cell Aperature on Mutation Response in *Tradescantia* Clone 4430 Following Exposure to Gaseous Ethylene Dibromide (DBE)

Mutagen	*Treatment*	*Exposure Rate*	*Mutant Events/Hair - Control ± S.E.*
DBE	Stomata open (exp. in light)	50.55 ppm—6 hrs.	0.03513 ± 0.00317
DBE	Stomata closed (exp. in dark)	50.82 ppm—6 hrs.	0.01551 ± 0.00137
Mutation increase attributable to open stomata			0.01962 ± 0.00345
			126 percent increase

TABLE 6.5
Differential Response of *Tradescantia* Clone 4430 to Two Different Methods of Treatment with Ethyl Methanesulfate (EMS): Vacuum Injection versus Nonvacuum Injection

Mutagen	*Treatment*	*Exposure Rate*	*Mutant Events/Hair - Control ± S.E.*
EMS	vac. inj. + gas flow	50 ppm—5 hrs.	0.19582 ± 0.01227
EMS	gas flow alone	50 ppm—5 hrs.	0.13281 ± 0.00841
Mutation increase attributable to vac. inj.			0.06301 ± 0.01487
			47 percent increase

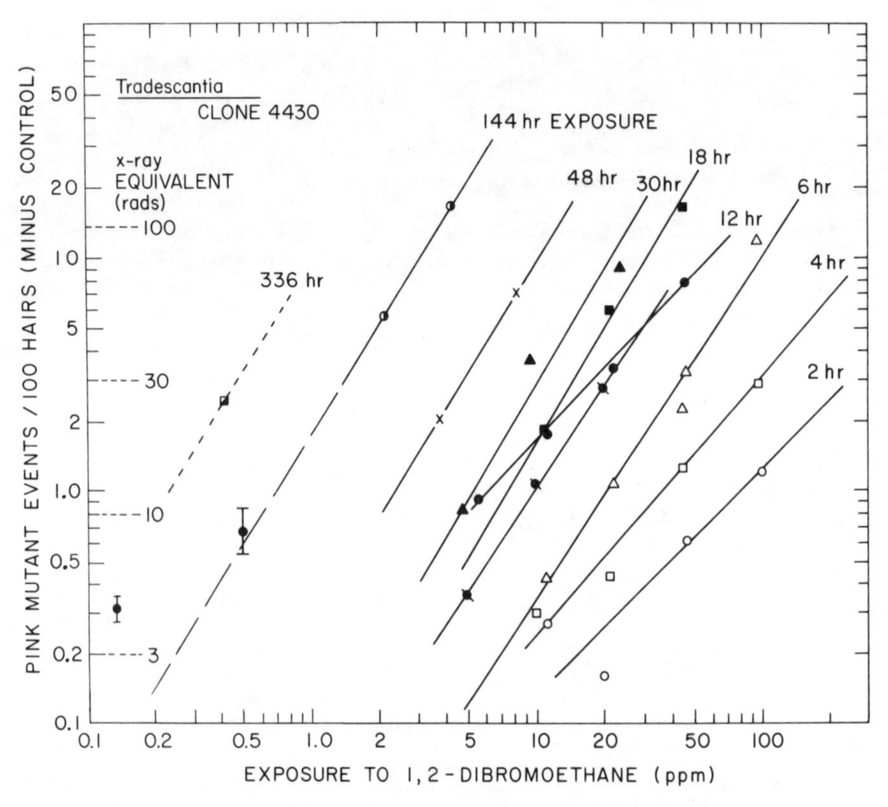

FIGURE 6.13. Pink Mutation Dose-Response Curves of *Tradescantia* Clone 4430 for Exposure Periods of 2 to 336 Hours and DBE Concentrations from 0.15 ppm to 100 ppm.

To define optimal gaseous exposure techniques further, the chemical mutagen 1,2-dibromoethane (DBE), an alkylating agent, was used as a standard. Flow-through exposures to DBE in the gaseous state were made for durations of 2 to 336 hours, and the resulting data show that mutational events increased directly with both increased exposure time and increased concentration (Figure 6.13). These data may also be expressed in terms of total dose by plotting induced mutation frequency against the product of concentration and duration of exposure (Figure 6.14). For purposes of comparison, a curve for X-ray effect is shown in rads, with radiation total dose defined as fluence × time. Slope and shape of the curve for DBE induction of color change resemble those for radiation-induced mutations.

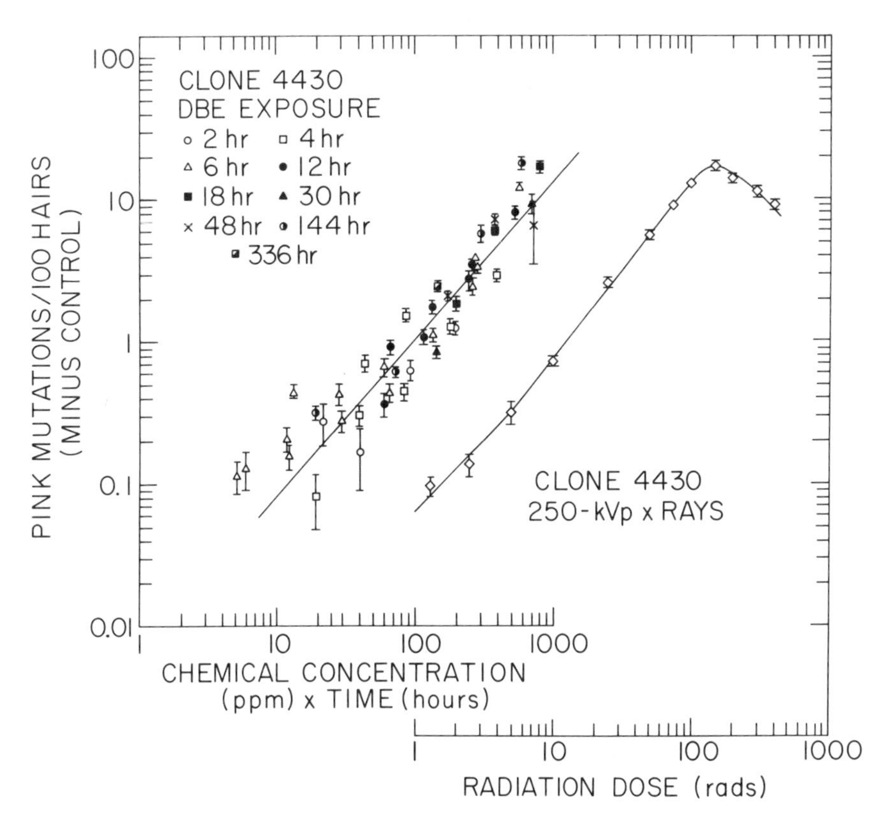

FIGURE 6.14 Pink Mutation Dose Response Plotted against Chemical Concentration (ppm × hrs). Note: The standard X ray curve is shown for comparison.

Although a larger percentage of the effort of this group has been spent on the development of the mobile monitoring vehicle, a number of chemicals have been tested in the laboratory to validate the system as a monitor for gaseous mutagens. Typical dose-response curves for several chemicals are shown in Figure 6.15. Chemicals such as ethyl methanesulfonate (EMS) and the gasoline additives 1,2-dibromoethane (DBE) and trimethylphosphate (TMP) were found to be potent mutagens, while SO_2, NO_2, vinyl chloride, and Freon-12 were weak mutagens according to this test system. Other chemicals or air pollutants tested in the vapor phase are listed in Table 6.6. The concentration listed is the lowest value tested that showed a significant mutagenic response.

Mutational events in low-dose experiments were confounded by seasonal variation in background mutation frequency, with a consistent cyclic high rate in the later summer months (Figure 6.16). The seasonal effect very likely was the result of ambient air pollution, since a clean-air chamber (one with a filtered atmosphere)

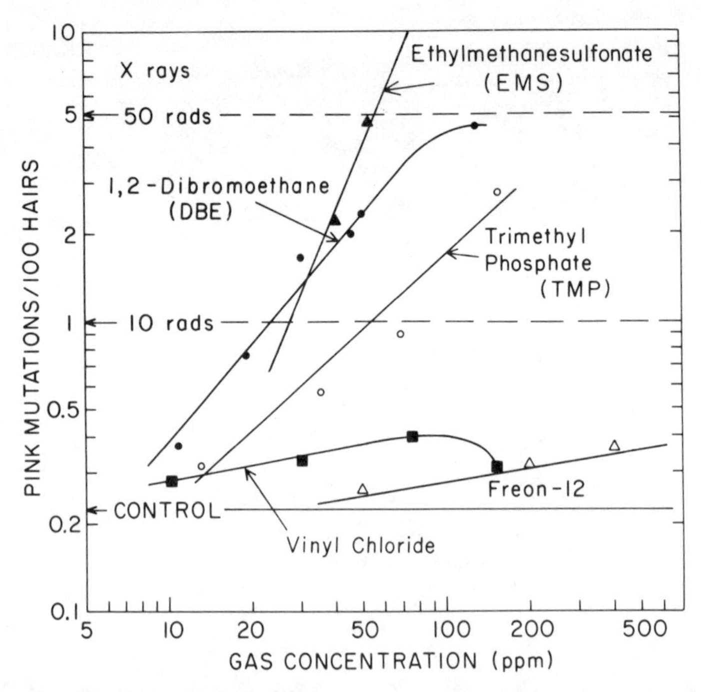

FIGURE 6.15. Dose-Response Curves for Vapor Phases of Ethyl Methanesulfonate, 1,2-Dibromoethane, Trimethyl Phosphate, Vinyl Chloride, and Freon-12.

TABLE 6.6
List of Gaseous Chemicals to Which *Tradescantia* Clone 4430 Was Exposed

Chemical	*Exposure Time (Hrs.)*	*Minimum* Conc.*	*Maximum** Conc.*	*Statistical Signif. (Percent)*
Ozone	6	5 ppm		2
Sulfur dioxide	6	40 ppm		1
Nitrogen dioxide	6	50 ppm		5
Nitrous oxide	6	250 ppm		1
Ethyl methane-sulfonate	6	5 ppm		1
1,2-dibromoethane	6	0.14 ppm		1
	144	0.14 ppm		1
Trimethyl-phosphate	13	2 ppm		2
Trichloroethylene	6	0.5 ppm		1
Vinyl chloride	6	75 ppm		2
Vinylidene	6	—	1,288 ppm	NS
chloride	24	22 ppm		5
Vinyl bromide	24	50 ppm		1
2-bromoethanol	6	24 ppm		2
Dichlorodifluoro-methane (Freon-12)	6	—	392 ppm	NS
Chlorodifluoro-methane (Freon-22)	6	194 ppm		2
Hexamethyl-phosphoramide	6	saturated		1
Benzene	6	4,000 ppm		1
1,1-dibromoethane	6	—	250 ppm	NS
Ethyl alcohol	24	1,000 ppm		1
Hydrozoic acid	6	3.5 ppm		1
Diallate	24	800 ppm		1
Tetrachloroethylene	6	1,035 ppm		5
Methylene chloride	6	15 ppm		1
1,1,1-trichloro-ethane	6	—	5,170 ppm	NS
Methane	72	—	50 percent	NS

Note: The mutagenic effect is given in terms of the level of statistical significance (percent).

*The lowest concentration at which an effect was observed.

**The highest concentration at which no effect was observed.

eliminated much of the background variability still seen in plants grown in a standard control chamber (Figure 6.17). Henceforth, filters of the type used in the clean-air chamber (activated charcoal and HEPA filters) were employed in stock plant chambers and in the mobile monitoring vehicle, to provide more stable baseline

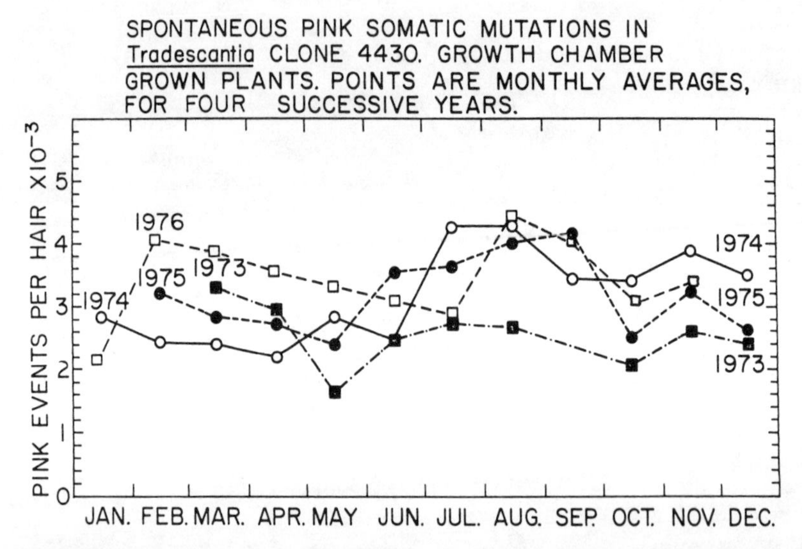

FIGURE 6.16. Seasonal Variation in Frequencies of Spontaneous Pink Events of Clone 4430 Grown in Controlled Environment Chambers. Note: Data points are monthly averages of control values from experiments over four successive years. Source: Schairer et al. (1978), with permission.

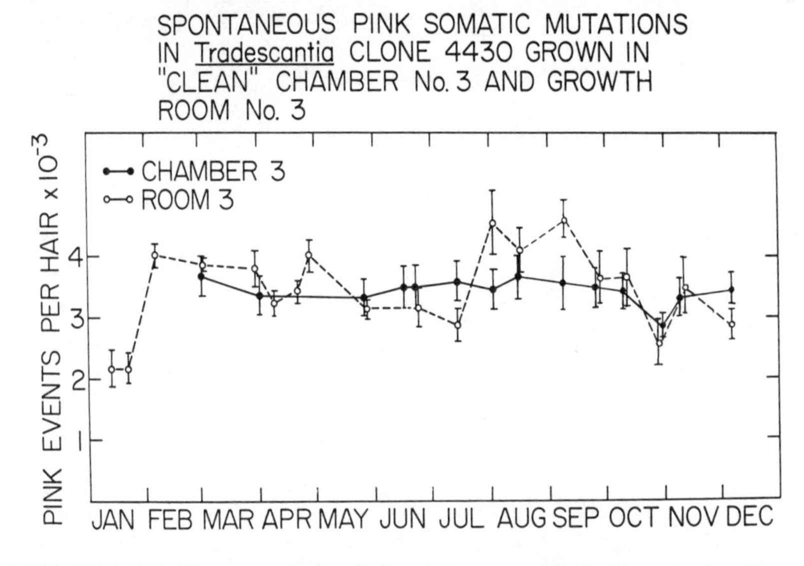

FIGURE 6.17. Frequencies of Spontaneous Pink Events in Stamen Hairs of *Tradescantia* Clone 4430 Plants Grown in a Standard Growth Chamber, Compared with Those from Plants Grown in "Clean" Chamber. Source: Schairer et al. (1978), with permission.

control data. The fact that the system responded to differences in filtered versus nonfiltered ambient air was further evidence for the high sensitivity of the color locus, and supported its use as a monitor for mutagens in ambient air.

MOBILE MONITORING VEHICLE (MMV) FOR IN SITU MONITORING OF AMBIENT AIR

Requirements for Field Exposures and Design of the MMV

The criteria for monitoring air pollution for mutagenicity include a roadworthy vehicle to house the test organism during exposures; exposure of the test organism under suitable culture conditions; a constant flow of untempered ambient air; and a semichronic exposure capability to simulate natural exposures to plants and animals. The vehicle selected for the mobile monitoring project was a 7.3-meter Clark mini-van trailer. The trailer, shown in Figure 6.18,

FIGURE 6.18 The Brookhaven Mobile Monitoring Vehicle (MMV), in an Operational Mode in the Field, with a Batelle Massive High-volume Sampler (right). The air-conditioning system for the trailer and the three growth chambers is on the front of the trailer. The meteorological monitoring equipment is on a pole above the trailer, and the glass ambient air intake ducts to the chambers are directly above the center of the trailer.

was insulated and air-conditioned to permit year-round operation of the laboratory. In order to maintain a semiclean environment for these studies, the trailer air was recirculated through activated charcoal and high efficiency particulate in air (HEPA) filters. Three model M-13 growth chambers (Environmental Growth Chambers, Chagrin Falls, Ohio) were installed.

One of the chambers serves as a clean-air control, the second is used for ambient air exposures, and the third is a back-up unit for either control or ambient air exposures (Figure 6.19). The chambers are designed to maintain any desired standard laboratory condition or to simulate fluctuations in the temperature and relative humidity of the ambient air outside. Ambient air is drawn into the fumigation chamber from a level 4.3 meters above ground through a 10-cm ID glass duct. Continuous flow rates up to about 0.51 cubic meters per minute are maintained to provide a maximum of one air change every two minutes. The clean-air control chamber is equipped with an air filter train composed of activated charcoal, HEPA particulate, and Purafil filters to scrub the air continuously while monitoring in situ controls. This procedure provides the cleanest air possible for background mutation rate determination.

Field Exposure Techniques

Field exposures are accomplished in the following manner: Fresh cuttings of *Tradescantia* clone 4430 are made from stock plants grown in controlled environment chambers at Brookhaven National Laboratory; they are randomly divided into sets of 60 cuttings, each set sealed in a polyethylene bag and hand-carried to the test site by car or plane; cuttings are placed randomly in the control and exposure chambers, in glass containers with Hoagland's nutrient solution (Figure 6.5); and exposures are made for 10-day periods. At the end of the exposure the cuttings are again sealed in plastic bags and returned to Brookhaven National Laboratory for posttreatment analysis of the flowers as they bloom each day. The peak mutation response period following a 10-day chronic exposure is 11 to 17 days after the start of the exposure. The mean of the mutation rates for the five- to seven-day scoring period results in an observed rate for a given test site, based on an average stamen hair population of between 150,000 and 300,000. A population of 300 cuttings in each ambient air and control chamber can yield enough data to resolve as small as a 10 percent increase in pink events over the background frequency.

FIGURE 6.19. The Interior of the MMV, with Two of the Three M-13 Growth Chambers. One door is open to show the *Tradescantia* cuttings, in an aerated nutrient solution, prepared for exposure. The humidifiers and filter train are on the front of each chamber. Source: Schairer et al. (1979), with permission.

Results of in Situ Exposures to Ambient Atmospheres

The first field trials for the MMV were conducted in the summer of 1976 at Elizabeth, New Jersey. Over the next four years, 17 additional sites throughout the United States were monitored in a preliminary survey. This study had two objectives: to demonstrate the adaptability of the stamen hair system to ambient air monitoring and, if mutagenicity was observed, to look for causative agents common to positive sites. These sites were selected because of known high levels of human cancer incidence or the presence of high levels of suspect compounds in the atmosphere. Two exceptions were the sites at Grand Canyon, Arizona, and Pittsboro, North Carolina. It was deemed essential to conduct a "clean ambient air" exposure to verify the efficiency of the filter on the concurrent control chamber and to rule out the possibility of an artifact being generated in the ambient air chamber. Two exposures at Grand Canyon produced similar results with no significant difference in background mutation frequency between the control and ambient air samples.

Since the Grand Canyon test represented a good baseline exposure to clean air in situ and included all of the stress of shipping plant material, field handling, and so on, the weighted mean for four replicated ambient air samples ($3.35 \pm 0.09/10^3$ hairs) was determined, and this mutation frequency was established as the standard for comparison for all other field sites monitored. In January 1980 the second clean air site, at Pittsboro, North Carolina, was monitored, and the results were in good agreement with those of the Grand Canyon study. The results of mutagenicity monitoring at all sites throughout the United States are shown in Figure 6.20, as the net induced mutation frequency following subtraction of the Grand Canyon standard control rate. Clearly significant increases in mutation frequency at the flower color locus were observed at many industrial sites. The most consistent mutagenic response was that associated with processing of petroleum products.

Mutation rate varied not only from site to site but also with repeated exposures at the same site. The induced mutation frequency ranged from 5.1 percent at Elizabeth, New Jersey, in April 1980 to 90.6 percent in September 1978. Some of this variation in response was undoubtedly due to seasonal change in pollution production and varying wind direction; prevailing winds in the summer and fall are from the southwest, while winter winds are from the northwest. The fixed location of the Elizabeth site placed petroleum refining operations directly upwind during the summer. It is important to note that atmospheric monitoring at fixed sites is very dependent

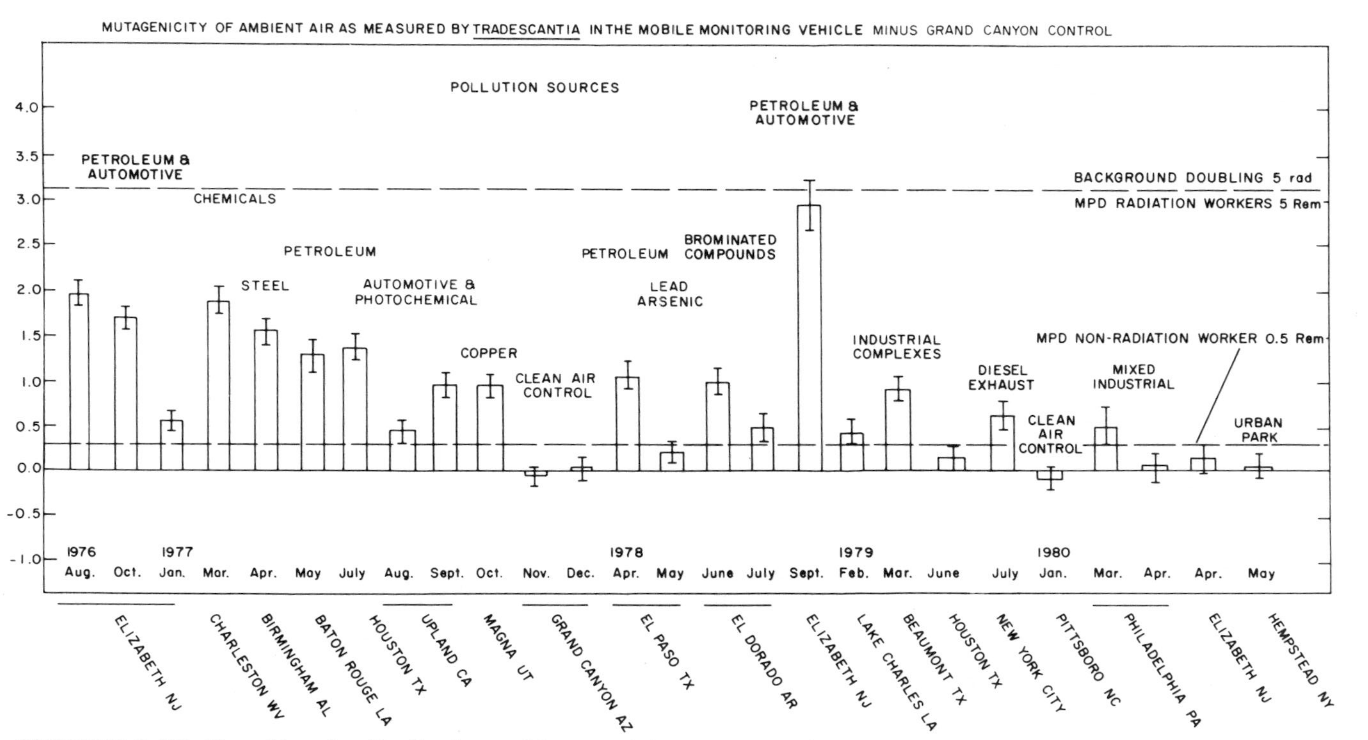

FIGURE 6.20. Results of a Preliminary Mutagenicity Monitoring Survey, Using *Tradescantia*, of 18 Sites Throughout the United States, Expressed as the Net Induced Mutation Frequency Following Subtraction of the Grand Canyon Standard Control Rate. Note: The broken horizontal lines indicate the level of mutation response in *Tradescantia* to radiation exposures at established maximum permissible dosages of 0.5 Rem for nonradiation workers and 5 Rem for radiation workers.

upon wind direction and speed. Although trailer locations for the 18 sites in this survey were selected downwind from the desired sources of pollution, a false negative may result from unpredicted wind changes, extended periods of rain, or other environmental factors.

SUMMARY AND CONCLUSIONS

Final assessment of human health effects resulting either directly or indirectly from exposure to harmful environmental agents may rest with mammalian test systems. In vitro systems, including mammalian cell and tissue cultures, are short-term assays used most frequently for extrapolation to humans. However, the present consensus is that no single assay system is adequate; discrepancies have been observed between false negatives, false positives, and metabolic activation of promutagens in established genetic systems such as *Salmonella*, *Saccharomyces*, *Drosophila*, and *Zea mays*. The more expensive long-term tests must be augmented by multiple assays designed for mere redundancy or to fill gaps uniquely in the present state of the art of environmental monitoring. The *Tradescantia* stamen hair test system is one such assay that can offer redundancy as well as fill the gap of monitoring ambient air for mutagenic agents.

The flower color locus in heterozygous clones of *Tradescantia* has been shown to mutate when exposed to various categories of agents ranging from fumigants (1,2-dibromoethane and trimethylphosphate) to solvents (hexamethylphosphoramide and trichloroethylene), to chemical additives or catalysts (vinyl chloride), and to compounds requiring activation, such as benzo(α)pyrene. This laboratory validation of the stamen hair system supports its use as a short-term bioassay for chemical mutagens.

The most unique application of this plant system is its ability to respond to low levels of airborne compounds. Without the constraints of sterile culture and other complicated experimental procedures, the *Tradescantia* plant is very adaptable to in situ monitoring at suspect industrial sites. Collection of data on the mutagenicity of ambient air pollution is an area of testing not amenable to mammalian or in vitro systems, and hence is an area in which *Tradescantia* can make its most valuable contribution.

The high sensitivity and versatility of the *Tradescantia* stamen hair system were put to good use by employing the plant as an in situ monitor for mutagens in ambient air at polluted industrial sites. Preliminary results from many sites showed a significant increase in mutation rate, with one (Elizabeth, New Jersey) approaching a background doubling. The atmosphere most consistently found to be mutagenic was that downwind from petroleum industrial complexes, although no specific compounds or groups of compounds have yet been correlated with the positive sites. Studies involving the fractionation and analysis of complex mixtures are continuing.

The relevance of the mutation increase attributed to exposure to ambient air pollution can be demonstrated by comparison with the radiation levels required to produce similar effects. Over 25 years of intensive radiobiological studies have generated health hazard exposure limits for industrial radiation workers of 5 Rem per year, and for nonradiation workers of 0.5 Rem per year. Figure 6.20 shows the pollution-induced mutation increment for the various sites monitored, and the broken horizontal lines indicate the mutation responses for 5 and 0.5 Rem, respectively. It is clear that one urban site (Elizabeth, New Jersey) showed a pollution response equal to that for the maximum permissible dose for a radiation worker, and most sites had a mutation response exceeding that permissible for nonradiation workers. The inference to be drawn from these data is that mutation frequency induced by environmental contamination from airborne chemicals alone is of a magnitude greater than that postulated as excessive for radiation exposure. Based on the responses of the *Tradescantia* stamen hair bioassay to in situ exposures to ambient air, many industrial sites warrant extensive investigation with the more expensive, long-term bioassays to determine the degree of human health hazard present.

FUTURE POSSIBILITIES

As an established short-term bioassay, the stamen hair system can be used as one of a battery or tier of tests for chemical mutagen assessment. The most unique contribution will be in the area of atmospheric monitoring. A collaborative study with Research Triangle Institute, under contract with the U.S. Environmental Protection Agency, has shown good correlation between mutagenicity of airborne particulates (Ames/*Salmonella*) and vapors (*Tradescantia*) at the same industrial sites (Hughes et al., 1980). Fractionation, quantitative analysis, and bioassessment of environmental mixtures, using the above and other organisms, may identify specific hazardous compounds that can be tested in the laboratory on *Tradescantia* and on the more rigorous mammalian systems. Fractionation techniques have the obvious advantages of helping to identify the pollution source and of simplifying atmospheric clean-up and regulation procedures for specific hazardous compounds.

Greater use of basic radiobiological data should be encouraged. There is a very large background of radiation data describing dose-response curve patterns (Nauman and Sparrow, 1978), dose rate effects (Underbrink and Sparrow, 1974; Nauman, Underbrink, and Sparrow, 1975), oxygen enhancement ratios in tissue (Underbrink et al., 1975), and nuclear factors influencing sensitivity (Sparrow et al., 1961). Dose-response curve patterns have been shown to be similar in *Tradescantia* for both physical and chemical agents. Other phenomena established by radiation studies should be explored for chemicals. Predictive potential for effects of environmental mutagens is as important as earlier studies related to nuclear fallout and its impact on ecology in general and the food chain in particular.

ACKNOWLEDGMENTS

This research, carried out at Brookhaven National Laboratory, was supported jointly by the U.S. Department of Energy and interagency agreements with the National Institute of Environmental Health Sciences and the U.S. Environmental Protection Agency. The authors acknowledge with thanks the special efforts of the Analytical Chemistry Division and of P. J. Klotz, in particular, for the development of the gaseous chemical exposure technique; N. R. Tempel, for engineering the assembly and subsequent operation of the mobile laboratory; and Dr. E. Pellizzari and staff of the Research Triangle Institute for organic vapor analysis. The many hours of flower analysis and genetic studies by E. E. Klug, A. F. Nauman, M. M. Nawrocky, V. Pond, R. C. Sparrow, and Dr. M. Emmerling-Thompson are also gratefully acknowledged.

REFERENCES

Attix, F. H. 1969. *Radiation Dosimetry*. 2nd ed. New York: Academic Press.

——. 1972. *Radiation Dosimetry*. Supp. 1. New York: Academic Press.

Bond, R. G., and C. P. Straub, eds. 1973. *Handbook of Environmental Control*. Vol. 1, *Air Pollution*. Cleveland: CRC Press.

Bottino, P. J., R. J. Bores, and A. H. Sparrow. 1973. Relative biological effectiveness of β and X irradiation for seedling growth and survival in barley and somatic mutations in *Tradescantia*. *Radiat. Res.* 55:602.

Brittin, W. E., ed. 1973. *Air and Water Pollution*. Boulder: Colorado Associated University Press.

Calbiochem-Behring Corp. *Technical Bulletin—Handling of Carcinogens and Hazardous Compounds.*

Commoner, B. 1971. *The Closing Circle; Nature, Man and Technology.* New York: Knopf.

Creech, J. L., Jr., and M. N. Johnson. 1974. Angiocarcinoma of the liver in the manufacture of polyvinylchloride. *J. Occup. Med.* 16:150.

Cuany, R. L., A. H. Sparrow, and V. Pond. 1958. Genetic response of *Antirrhinum majus* to acute and chronic plant irradiation. *Z. indukt. Abstamm Vererbungsl.* 89:7-13.

Delone, N. L., A. S. Trusova, E. M. Morozova, V. Y. Ammunov, and G. P. Parfenov. 1968. Effects of space flight of the Sputnik Cosmos-110 on the microspores of *Tradescantia paludosa. Kosmicheskiye Issled.* 6:299-303.

de Serres, F. J., and M. D. Shelby, eds. 1978. Higher plant systems as monitors of environmental mutagens. Proceedings of NIEHS-sponsored workshop, Marineland, Florida, January 15-19, 1978. *Environ. Health Persp.* 27:1-206.

Eisenbud, M. 1973. *Environmental Radioactivity.* New York: Academic Press.

Epstein, S. S. 1974. Environmental determinants of human cancer. *Cancer Res.* 34:2425-2435.

Epstein, S. S., and M. S. Legator, eds. 1971. *The Mutagenicity of Pesticides. Concepts and Evaluation.* Cambridge, Mass.: MIT Press.

Fishbein, L. 1973. Mutagens and potential mutagens in the biosphere. *Mutat. Res.* 21:220.

Fishbein, L., W. G. Flamm, and H. L. Falk. 1970. *Chemical Mutagens, Environmental Effects on Biological Systems.* New York: Academic Press.

Flamm, W. G. 1974. A tier system approach to mutagen testing. *Mutat. Res.* 26:329.

Flamm, W. G., and M. A. Mehlman, eds. 1978. *Advances in Modern Toxicology.* Vol. 5, *Mutagenesis.* New York: John Wiley and Sons.

Fowler, E. B., ed. 1965. *Radioactive Fallout, Soils, Plants, Foods, Man.* New York: Elsevier.

Heath, C. W., Jr. 1978. Environmental pollutants and the epidemiology of cancer. *Environ. Health Persp.* 27:7-10.

Higginson, J. 1968. Present trends in cancer epidemiology. In *Proceedings of 8th Canadian Cancer Conference*, pp. 40-75. Honey Harbour, 1968. New York: Pergamon Press.

Hindawi, I. J. 1970. *Air Pollution Injury to Vegetation.* National Air Pollution Control Administration Publ. no. AP-71. Washington, D.C.: U.S. Dept. of Health, Education and Welfare.

Hollaender, A., and F. J. de Serres, eds. 1971-1980. *Chemical Mutagens: Principles and Methods for Their Detection.* 6 vols. New York: Plenum Press.

Hughes, T. J., E. Pellizzari, L. Little, C. Sparacino, and A. Kolber. 1980. Ambient air pollution: Collection, chemical characterization and mutagenicity testing. *Mutat. Res.* 76:51-83.

Klekowski, E. J., Jr. 1978. Detection of mutational damage in fern populations: An in situ bioassay for mutagens in aquatic ecosystems. In *Chemical*

Mutagens: Principles and Methods for Their Detection. Vol. 5, ed. A. Hollaender and F. de Serres, pp. 79-99. New York: Plenum Press.

Knoll, G. F. 1979. *Radiation Detection and Measurement.* New York: Wiley.

Kudirka, D. T., and J. van't Hof. 1980. G_2 arrest and differentiation in the petal of *Tradescantia* clone 4430. *Exptl. Cell Res.* 130:443-450.

Kusnetz, H. L., B. E. Saltzman, and M. E. Lanier. 1960. Calibration and evaluation of gas detecting tubes. *Am. Ind. Hyg. Assoc. J.* 21:361-373.

Larsen, P. A., R. A. Conrad, K. Knudsen, J. Robbins, J. Wolff, J. E. Rall, and B. Dobyns. 1978. Thyroid hypofunction appearing as a delayed manifestation of accidental exposure to radioactive fallout in a Marshallese population. Paper presented at IAEA meeting, Vienna, March 1978.

McNulty, P. J., A. H. Sparrow, and L. A. Schairer. 1974. Somatic mutations induced in *Tradescantia* clone 02 stamen hairs by relativistic muons. *Intern. J. Radiat. Biol.* 25:315-320.

Mericle, L. W., and R. P. Mericle. 1965. Biological discrimination of differences in natural background radiation level. *Radiat. Bot.* 5:475-492.

——. 1967. Genetic nature of somatic mutations for flower color in *Tradescantia*, clone 02. *Radiat. Bot.* 7:449-464.

Muller, H. J. 1927. Artificial transmutation of the gene. *Science* 66:84-87.

National Council on Radiation Protection and Measurement. 1980. *Influence of Dose and Its Distribution in Time on Dose-Response Relationships for Low-LET Radiations.* NCRP Report no. 64. Washington, D.C.: NCRP.

Nauman, C. H., P. J. Klotz, and L. A. Schairer. 1979. Uptake of tritiated 1,2-dibromoethane by *Tradescantia* floral tissues: Relation to induced mutation frequency in stamen hair cells. *Environ. Exp. Bot.* 19:201-215.

Nauman, C. H., and A. H. Sparrow. 1978. Problems of extrapolation from high dose to low dose in *Tradescantia* mutation studies. *Environ. Health Persp.* 22:161-162.

Nauman, C. H., A. H. Sparrow, and L. A. Schairer. 1976. Comparative effects of ionizing radiation and two gaseous chemical mutagens on somatic mutation induction in one mutable and two non-mutable clones of *Tradescantia. Mutat. Res.* 38:53-70.

Nauman, C. H., A. G. Underbrink, and A. H. Sparrow. 1975. Influence of radiation dose rate on somatic mutation induction in *Tradescantia* stamen hairs. *Radiat. Res.* 62:79-96.

Neel, J. V. 1974. Developments in monitoring human populations for mutation rates. *Mutat. Res.* 26:319-328.

Norman, J. H., and P. Winchill. 1971. Physical, chemical, and radiological properties of fallout. In *Survival of Food Crops and Livestock in the Event of a Nuclear War.* AEC Symposium no. 24, ed. D. W. Benson and A. H. Sparrow, pp. 9-30. Oak Ridge, Tenn.: U.S. Atomic Energy Commission.

Oser, B. L. 1971. Toxicology of pesticides to establish proof of safety. In *Pesticides in the Environment.* Vol. 1, Pt. II, ed. R. White-Stevens, pp. 411-456. New York: Marcel Dekker.

Russell, R. S., B. O. Bartlett, and R. S. Bruce. 1971. The significance of long-lived nuclides after a nuclear war. In *Survival of Food Crops and Livestock in the Event of a Nuclear War.* AEC Symposium no. 24, ed. D. W. Benson and A. H. Sparrow, pp. 548-565. Oak Ridge, Tenn.: U.S. Atomic Energy Commission.

Savage, J. R. K. 1975. Radiation-induced chromosomal aberrations in the plant *Tradescantia:* Dose-response curves. I. Preliminary considerations. *Radiat. Bot.* 15:87-140.

Schairer, L. A., J. Van't Hof, C. G. Hayes, R. M. Burton, and F. J. de Serres. 1978. Exploratory monitoring of air pollutants for mutagenicity activity with the *Tradescantia* stamen hair system. *Environ. Health Perspect.* 27:51-60.

Schairer, L. A., J. Van't Hof, C. G. Hayes, R. M. Burton, and F. J. de Serres. 1979. Measurement of biological activity of ambient air mixtures using a mobile laboratory for *in situ* exposures: Preliminary results from the *Tradescantia* plant test system. In *Application of Short-Term Bioassays in the Fractionation of Complex Environmental Mixtures,* ed. M. Waters, S. Nesnow, J. Huisingh, S. Sandhu, and L. Claxton, pp. 421-440. New York: Plenum Press.

Singleton, W. R., ed. 1958. *Nuclear Radiation in Food and Agriculture.* Princeton, N.J.: D. Van Nostrand.

Skelley, J. M. 1980. Photochemical oxidant impact on Mediterranean and temperate forest ecosystems: Real and potential effects. In *Proceedings of Symposium on Effects of Air Pollutants on Mediterranean and Temperate Forest Ecosystems, Pacific Southwest Forest and Range Experiment Station,* pp. 38-50.

Sparrow, A. H. 1951. Radiation sensitivity of cells during mitotic and meiotic cycles with emphasis on possible cytochemical changes. *Ann. N.Y. Acad. Sci.* 51:1508-1540.

——. 1960. Uses of large sources of ionizing radiation in botanical research and some possible practical applications. In *Large Radiation Sources in Industry.* Vol. 2, pp. 195-219. Vienna: IAEA.

——. 1962. The role of the cell nucleus in determining radiosensitivity. Brookhaven Lecture Series 17, BNL-766(T-287). Upton, N.Y.: Brookhaven National Laboratory.

Sparrow, A. H., K. P. Bactke, D. L. Shaver, and V. Pond. 1968. The relationship of mutation rate per roentgen to DNA content per chromosome and to interphase chromosome volume. *Genetics* 59:65-78.

Sparrow, A. H., R. L. Cuany, J. P. Miksche, and L. A. Schairer. 1961. Some factors affecting the responses of plants to acute and chronic radiation exposures. *Radiat. Bot.* 1:10-34.

Sparrow, A. H., and J. E. Gunckel. 1956. The effects on plants of chronic exposure to gamma radiation from radiocobalt. *Proceedings of the International Conference on Peaceful Uses of Atomic Energy.* Vol. 12, pp. 52-59.

Sparrow, A. H., and V. Pond. 1956. The relationship between dose rate and somatic mutation in *Antirrhinum majus* exposed to chronic gamma irradiation. *Radiat. Res.* 5:596.

Sparrow, A. H., L. A. Schairer, and K. M. Marimuthu. 1968. Genetic and cytologic studies of *Tradescantia* irradiated during orbital flight. *BioScience* 18:582-590.

——. 1971. Radiobiologic studies of *Tradescantia* plants orbited in Biosatellite II. In *The Experiments of Biosatellite II*, ed. J. F. Saunders, pp. 99-122. NASA SP-204. Washington, D.C.: NASA.

Sparrow, A. H., L. A. Schairer, M. M. Nawrocky, and R. C. Sautkulis. 1971. Effects of low temperature and low level chronic gamma radiation on somatic mutation rates in *Tradescantia*. *Radiat. Res.* 47:273-274.

Sparrow, A. H., L. A. Schairer, and R. C. Sparrow. 1963. Relationship between nuclear volumes, chromosome numbers, and relative radiosensitivities. *Science* 141:163-166.

Sparrow, A. H., L. A. Schairer, and R. Villalobos-Pietrini. 1974. Comparison of somatic mutation rates induced in *Tradescantia* by chemical and physical mutagens. *Mutat. Res.* 26:265-276.

Sparrow, A. H., and W. R. Singleton. 1953. The use of radiocobalt as a source of gamma rays and some effects of chronic irradiation on growing plants. *Am. Nat.* 87:29-48.

Sparrow, A. H., A. G. Underbrink, and H. H. Rossi. 1972. Mutations induced in *Tradescantia* by small doses of x rays and neutrons: Analysis of dose-response curves. *Science* 176:916-918.

Stadler, L. J. 1928. Genetic effects of x rays in maize. *Proc. Natl. Acad. Sci. U.S.A.* 14:69-75.

Thompson-Emmerling, M., and M. M. Nawrocky. 1980. Genetic basis for using *Tradescantia* clone 4430 as an environmental monitor of mutagens. *J. Hered.* 71:261-265.

Underbrink, A. G., L. A. Schairer, and A. H. Sparrow. 1973a. The biophysical properties of 3.9-GeV nitrogen ions. V. Determination of the relative biological effectiveness for somatic mutations in *Tradescantia*. *Radiat. Res.* 55:437-446.

——. 1973b. *Tradescantia* stamen hairs: A radiobiological test system applicable to chemical mutagenesis. In *Chemical Mutagens: Principles and Methods for Their Detection.* Vol. 3, ed. A. Hollaender, pp. 171-207. New York: Plenum Press.

Underbrink, A. G., and A. H. Sparrow. 1974. The influence of experimental end points, dose, dose rate, neutron energy, nitrogen ions, hypoxia, chromosome volume and ploidy level on RBE in *Tradescantia* stamen hairs and pollen. In *Biological Effects of Neutron Irradiation.* Vienna: International Atomic Energy Agency.

Underbrink, A. G., A. H. Sparrow, D. Sautkulis, and R. E. Mills. 1975. Oxygen enhancement ratios (OER) for somatic mutations in *Tradescantia* stamen hairs. *Radiat. Bot.* 15:161-168.

Vogel, F., and G. Röhrborn, eds. 1970. *Chemical Mutagenesis in Mammals and Man.* New York: Springer Verlag.

White-Stevens, R., ed. 1971. *Pesticides in the Environment.* Vol. 1, Pt. II. New York: Marcel Dekker.

Index